BIOCHEMISTRY

BIOCHEMISTRY

Edited by
N.P. SINGH

AKANSHA PUBLISHING HOUSE
NEW DELHI - 110 002 (INDIA)

AKANSHA PUBLISHING HOUSE
R-37B, Vani Vihar, Uttam Nagar
New Delhi-110 059
E-mail: ektabooks@yahoo.com

Showroom:
4649/21B, Ansari Road,
Daryaganj, New Delhi-110 002
Phone: 9212065579, 9811582579

Biochemistry

First Published, 2005

ISBN 81-87606-79-7

PRINTED IN INDIA

Published by M.P. Misra for Akansha Publishing House, New Delhi-110 002 and Printed at Tarun Offset Press, Delhi.

Preface

From the smallest microbe to the largest mammal, all life depends on the properties and reactions four classes of organic compounds—carbohydrates, proteins, lipids, and nucleic acids. Carbohydrates provide energy for metabolism. Proteins catalyse the reactions that allow an organism to live and grow. Lipids build cell membranes that regulate the internal structure and function of cells. Nucleic acids supply the blueprint for the synthesis of proteins and the function of life itself. Biochemistry is the study of the structure, properties, and reactions of these compounds.

Throughout the history of this scientific discipline, biochemists have worked to reveal the fundamental chemical and physical principles that underlie living processes. Their success is demonstrated in the enormous impact that the biochemical approach has had on the life sciences. At the beginning of the twenty-first century, the depth and breadth of this influence is astonishing.

The progress in our understanding of living organisms, already enormous by the early 1990s, is currently being exceeded, in no small measure, because of the spectacular advances in DNA-based technologies that are based on biochemical research. The access that life scientists now have to the genetic information of entire organisms has resulted in previously unimaginable insights into the inner workings of living organisms and the causes of disease. The challenge in life and physical science; education is how to prepare students for careers in diverse fields in which the pace of knowledge accumulation will only accelerate in the foreseeable future. The most important tool that teachers can offer these students is a coherent understanding of biochemistry.

This book is designed for life science students and chemistry and other physical science majors. Few assumptions have been made about the chemistry and biology backgrounds the students have. Emphasis is placed on the relationship between moleculai architecture and the functional properties of biomolecules, and the dynamic, unceasing, and self-regulating nature of living processes. Students are also provided with overviews of the major physical and chemical techniques that biochemists have used to explore life at the molecular level.

Editor

Contents

1

Basics of Biochemistry

Biochemistry asks how the remarkable properties of living organisms arise from the thousands of different lifeless biomolecules. When these molecules are isolated and examined individually, they conform to all the physical and chemical laws that describe the behaviour of inanimate matter-as do all the processes occurring in living organisms. The study of biochemistry shows how the collections of inanimate molecules that constitute living organisms interact to maintain and perpetuate life animated solely by the physical and chemical laws that govern the nonliving universe.

Yet organisms possess extraordinary attributes, properties that distinguish them from other collections of matter. Following are the distinguishing features of living organisms:

- A high degree of chemical complexity and microscopic organization. Thousands of different molecules make up a cell's intricate internal structures. Each has its characteristic sequence of subunits, its unique three-dimensional structure, and its highly specific selection of binding partners in the cell.

— Systems for extracting, transforming, and using energy from the environment , enabling organisms to build and maintain their intricate structures and to do mechanical, chemical, osmotic, and electrical work. Inanimate matter tends, rather, to decay toward a more disordered state, to come to equilibrium with its surroundings.

— A capacity for precise self-replication and self-assembly . A single bacterial cell placed in a sterile nutrient medium can give rise to a billion identical "daughter" cells in 24 hours. Each cell contains thousands of different molecules, some extremely complex; yet each bacterium is a faithful copy of the original, its construction directed entirely from information contained within the genetic material of the original cell.

— Mechanisms for sensing and responding to alterations in their surroundings, constantly adjusting to these changes by adapting their internal chemistry.

— Defined functions for each of their components and regulated interactions among them. This is true not only of macroscopic structures, such as leaves and stems or hearts and lungs, but also of microscopic intracellular structures and individual chemical compounds. The interplay among the chemical components of a living organism is dynamic; changes in one component cause coordinating or compensating changes in another, with the whole ensemble displaying a character beyond that of its individual parts. The collection of molecules carries out a program, the end result of which is reproduction of the program and self-perpetuation of that collection of molecules-in short, life.

— A history of evolutionary change. Organisms change their inherited life strategies to survive in new circumstances. The result of eons of evolution is an enormous diversity of life forms, superficially very different but fundamentally related through their shared ancestry.

Despite these common properties, and the fundamental unity of life they reveal, very few generalizations about living organisms are absolutely correct for every organism under every condition; there is enormous diversity. The range of habitats in which organisms live, from hot springs to Arctic tundra, from animal intestines to college dormitories, is matched by a correspondingly wide range of specific biochemical adaptations, achieved within a common chemical framework. For the sake of clarity, in this book we sometimes risk certain generalizations, which, though not perfect, remain useful; we also frequently point out the exceptions that illuminate scientific generalizations.

Biochemistry describes in molecular terms the structures, mechanisms, and chemical processes shared by all organisms and provides organizing principles that underlie life in all its diverse forms, principles we refer to collectively as the molecular logic of life. Although biochemistry provides important insights and practical applications in medicine, agriculture, nutrition, and industry, its ultimate concern is with the wonder of life itself.

Cellular Foundations

The unity and diversity of organisms become apparent even at the cellular level. The smallest organisms consist of single cells and are microscopic. Larger, multicellular organisms contain many different types of cells, which vary in size, shape, and specialized function. Despite these obvious differences, all cells of the simplest and most complex organisms share certain fundamental properties, which can be seen at the biochemical level.

Functional Units of Living Organisms

Cells of all kinds share certain structural features (Fig. 1). The plasma membrane defines the periphery of the cell, separating its contents from the surroundings. It is composed of lipid and protein molecules that form a thin, tough, pliable, hydrophobic barrier around the cell. The membrane is a barrier to the free passage of inorganic ions and most other charged or polar compounds. Transport proteins in the plasma membrane allow the passage of certain ions and molecules; receptor proteins transmit signals into the cell; and membrane enzymes participate in some reaction pathways. Because the individual lipids and proteins of the plasma membrane are not covalently linked, the entire structure is remarkably flexible, allowing changes in the shape and size of the cell. As a cell grows, newly made lipid and protein molecules are inserted into its plasma membrane; cell division produces two cells, each with its own membrane. This growth and cell division (fission) occurs without loss of membrane integrity.

The internal volume bounded by the plasma membrane, the cytoplasm (Fig. 1), is composed of an aqueous solution, the cytosol, and a variety of suspended particles with specific functions. The cytosol is a highly concentrated solution containing enzymes and the RNA molecules that encode them; the components (amino acids and nucleotides) from which these macro-molecules are assembled; hundreds of small organic molecules called metabolites, intermediates in biosynthetic and degradative pathways; coenzymes, compounds essential to many enzyme catalyzed reactions; inorganic ions; and ribosomes, small particles (composed of protein and RNA molecules) that are the sites of protein synthesis.

All cells have, for at least some part of their life, either a nucleus or a nucleoid, in which the genome-the complete set of genes, composed of DNA-is stored and replicated. The nucleoid, in bacteria, is not separated from the cytoplasm by a membrane; the nucleus, in higher organisms, consists of nuclear material enclosed within a double membrane, the nuclear envelope. Cells with nuclear envelopes are called eukaryotes (Greek eu, "true," and karyon, "nucleus"); those without nuclear envelopes-bacterial cells-are prokaryotes (Greek pro, "before").

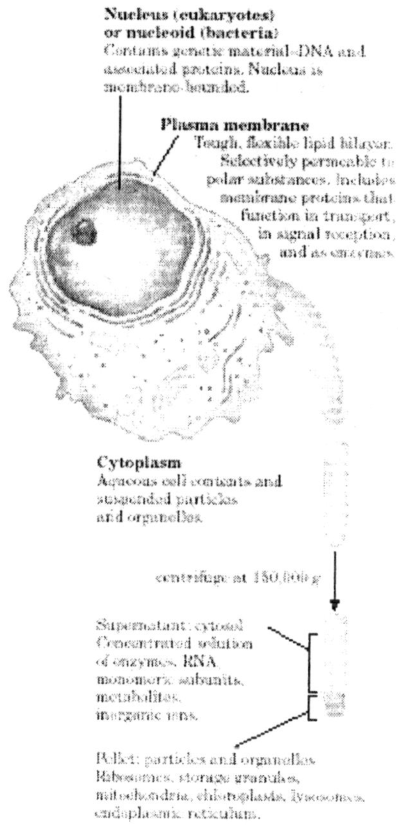

Figure 1: The universal features of living cells.

Cellular dimensions are limited by oxygen diffusion

Most cells are microscopic, invisible to the unaided eye. Animal and plant cells are typically 5 to 100 μm in diameter, and many bacteria are only 1 to 2 μm long . What limits the dimensions of a cell? The lower limit is probably set by the minimum number of each type of biomolecule required by the cell. The smallest cells, certain bacteria known as mycoplasmas, are 300 nm in diameter and have a volume of about 10^{-14} mL. A single bacterial ribosome is about 20 nm in its longest dimension, so a few ribosomes take up a substantial fraction of the volume in a mycoplasmal cell.

The upper limit of cell size is probably set by the rate of diffusion of solute molecules in aqueous systems. For example, a bacterial cell that depends upon oxygen-consuming reactions for energy production must obtain molecular oxygen by diffusion from the surrounding medium through its plasma membrane. The cell is so small, and the ratio of its surface area to its volume is so large, that every part of its cytoplasm is easily reached by O_2 diffusing into the cell. As cell size increases, however, surface-to-volume ratio decreases, until metabolism consumes O_2 faster than diffusion can supply it. Metabolism that requires O_2 thus becomes impossible as cell size increases beyond a certain point, placing a theoretical upper limit on the size of the cell.

Domains of Life

All living organisms fall into one of three large groups (kingdoms, or domains) that define three branches of evolution from a common progenitor (Fig. 2). Two large groups of prokaryotes can be distinguished on biochemical grounds: archaebacteria (Greek arche - , "origin") and eubacteria (again, from Greek eu, "true"). Eubacteria inhabit soils, surface waters, and the tissues of other living or decaying organisms. Most of the well-studied bacteria, including Escherichia coli, are eubacteria. The archaebacteria, more recently discovered, are less well characterized biochemically; most inhabit extreme environments-salt lakes, hot springs, highly acidic bogs, and the ocean depths. The available evidence suggests that the archaebacteria and eubacteria diverged early in evolution and constitute two separate domains, sometimes called Archaea and

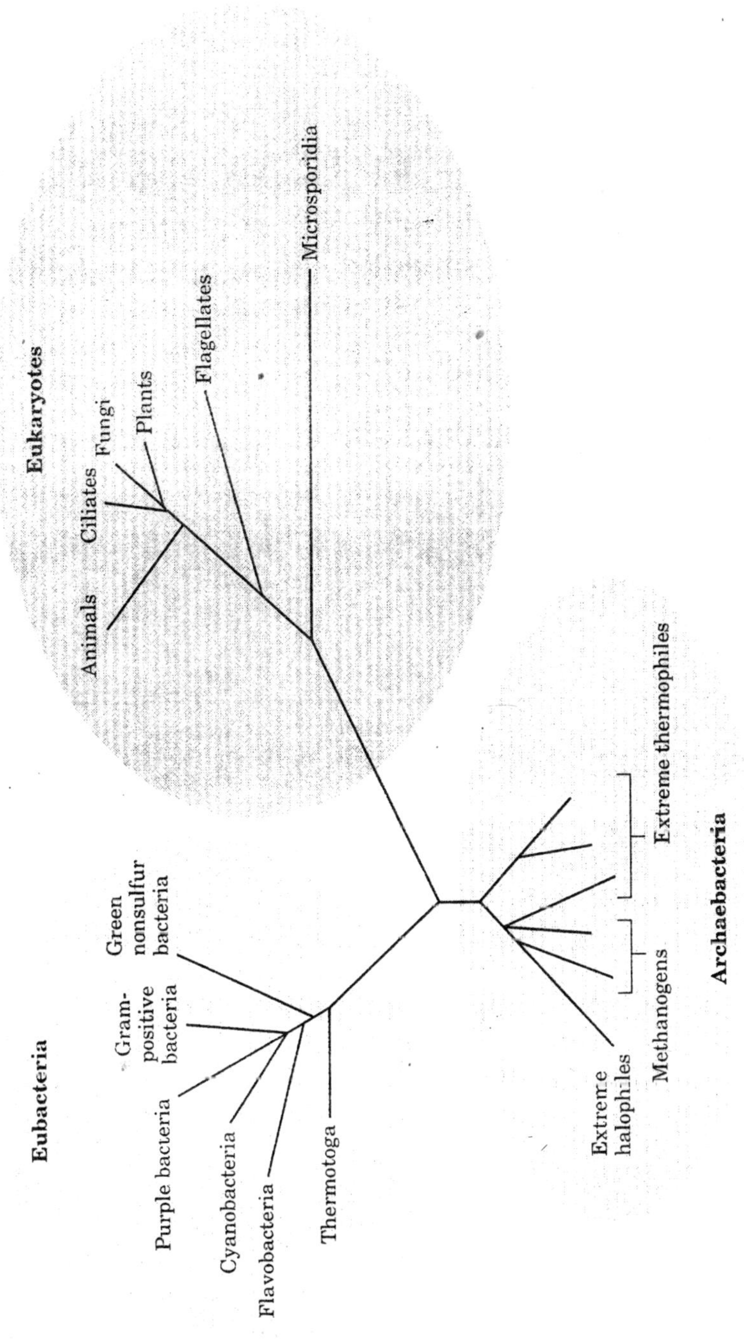

Figure 2: Phylogeny of the three domains of life.

Bacteria. All eu-karyotic organisms, which make up the third domain, Eukarya, evolved from the same branch that gave rise to the Archaea; archaebacteria are therefore more closely related to eukaryotes than to eubacteria.

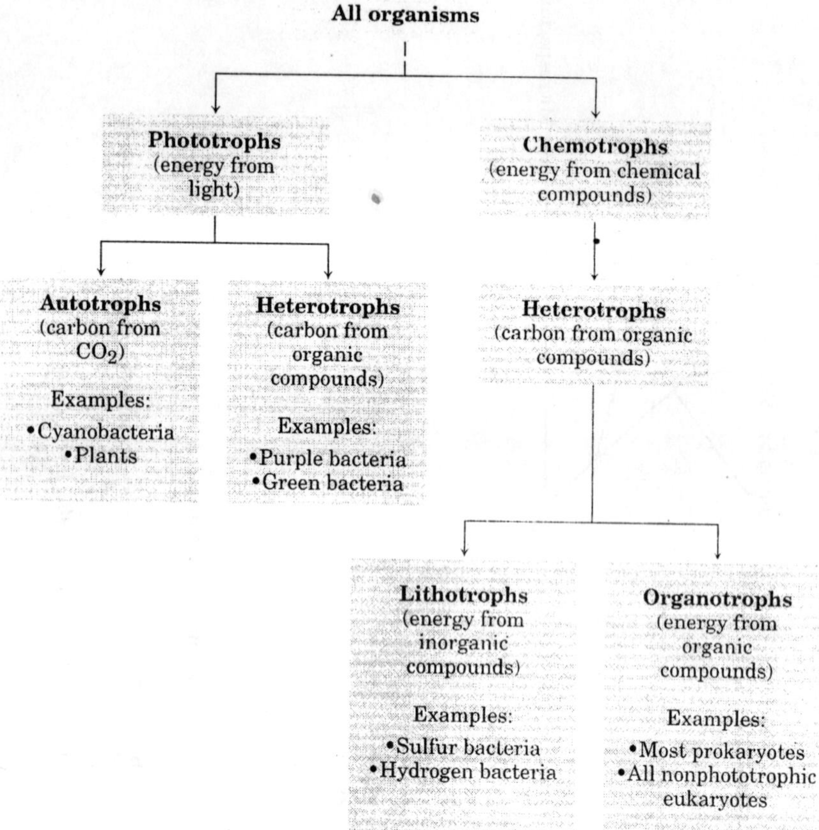

Figure 3: Classification of Organisms

Within the domains of Archaea and Bacteria are subgroups distinguished by the habitats in which they live. In aerobic habitats with a plentiful supply of oxygen, some resident organisms derive energy from the transfer of electrons from fuel molecules to oxygen. Other environments are anaerobic, virtually devoid of oxygen, and microorganisms adapted to these environments obtain energy by transferring electrons to nitrate (forming N_2), sulfate (forming H_2S), or CO_2 (forming CH_4). Many organisms that have evolved in anaerobic environments are obligate anaerobes: they die when exposed to oxygen.

We can classify organisms according to how they obtain the energy and carbon they need for synthesizing cellular material. There are two broad categories based on energy sources: phototrophs (Greek trophe, "nourishment") trap and use sunlight, and chemotrophs derive their energy from oxidation of a fuel. All chemotrophs require a source of organic nutrients; they cannot fix CO_2 into organic compounds. The phototrophs can be further divided into those that can obtain all needed carbon from CO_2 (autographs) and those that require organic nutrients (heterotrophs). No chemotroph can get its carbon atoms exclusively from CO_2 (that is, no chemotrophs are autotrophs), but the chemotrophs may be further classified according to a different criterion: whether the fuels they oxidize are inorganic (lithotrophs) or organic (organotrophs).

Most known organisms fall within one of these four broad categories-autotrophs or heterotrophs among the photosynthesizers, lithotrophs or organotrophs among the chemical oxidizers. The prokaryotes have several general modes of obtaining carbon and energy. Escherichia coli, for example, is a chemoorganoheterotroph; it requires organic compounds from its environment as fuel and as a source of carbon. Cyanobacteria are photo-lithoautotrophs; they use sunlight as an energy source and convert CO_2 into biomolecules. We humans, like *E.* coli, are chemoorganoheterotrophs.

Escherichia coli

Bacterial cells share certain common structural features, but also show group-specific specializations. *E.*coli is a usually harmless inhabitant of the human intestinal tract. The coli cell is about 2 mm long and a little less than 1 mm in diameter. It has a protective outer membrane and an inner plasma membrane that

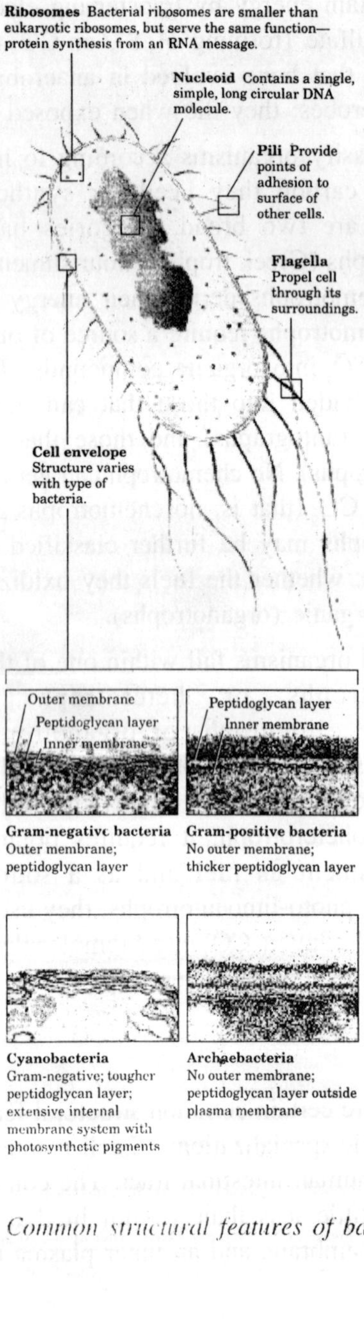

Figure 4: Common structural features of bacterial cells.

encloses the cytoplasm and the nucleoid. Between the inner and outer membranes is a thin but strong layer of polymers called peptidoglycans, which gives the cell its shape and rigidity. The plasma membrane and the layers outside it constitute the cell envelope. In the Archaea, rigidity is conferred by a different type of polymer (pseudopeptidoglycan). The plasma membranes of eubacteria consist of a thin bilayer of lipid molecules penetrated by proteins. Archaebacterial membranes have a similar architecture, although their lipids differ strikingly from those of the eubacteria.

The cytoplasm of *E.* coli contains about 15,000 ribosomes, thousands of copies each of about 1,000 different enzymes, numerous metabolites and co-factors, and a variety of inorganic ions. The nucleoid contains a single, circular molecule of DNA, and the cytoplasm contains one or more smaller, circular segments of DNA called plas-mids. In nature, some plasmids confer resistance to toxins and antibiotics in the environment. In the laboratory, these DNA segments are especially amenable to experimental manipulation and are extremely useful to molecular geneticists.

Most bacteria (including *E.* coli) lead existences as individual cells, but in some bacterial species cells tend to associate in clusters or filaments, and a few (the myxobacteria, for example) demonstrate simple social behaviour.

Membranous Organelles

Typical eukaryotic cells are much larger than prokaryotic cells-commonly 5 to 100 μm in diameter, with cell volumes a thousand to a million times larger than those of bacteria. The distinguishing characteristics of eukaryotes are the nucleus and a variety of membrane-bounded organelles with specific functions: mitochondria, endoplasmic reticulum, Golgi complexes, and lysosomes. Plant cells also contain vacuoles and chloroplasts . Also present in the cytoplasm of many cells are granules or droplets containing stored nutrients such as starch and fat.

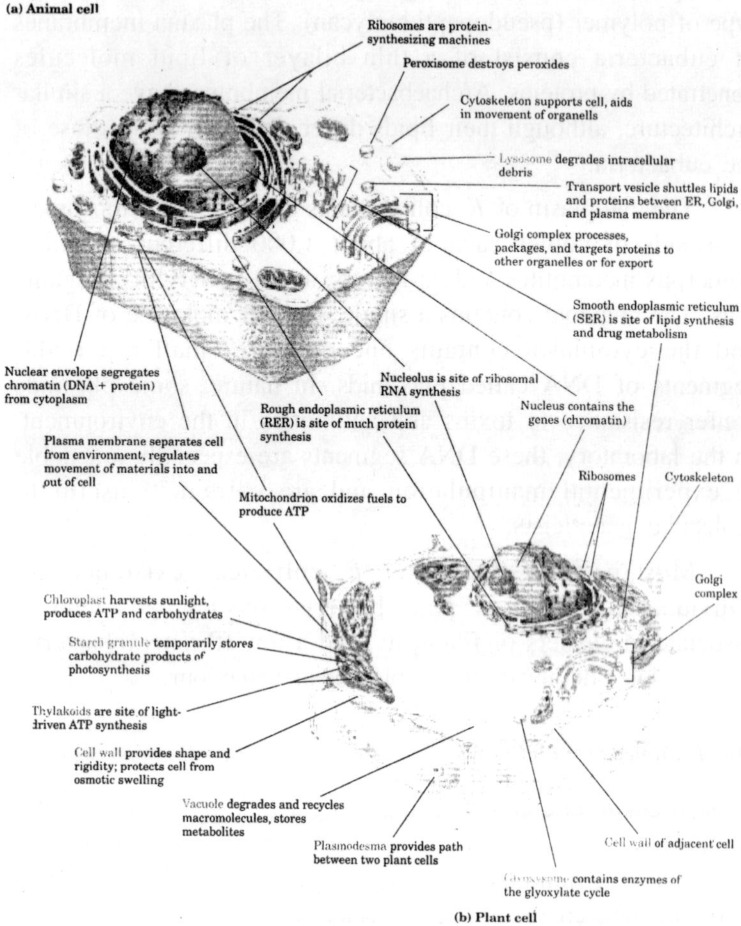

Figure 5: Eukaryotic cell structure

In a major advance in biochemistry, Albert Claude, Christian de Duve, and George Palade developed methods for separating organelles from the cytosol and from each other-an essential step in isolating biomolecules and larger cell components and investigating their structures and functions. In a typical cell fractionation (Fig. 6), cells or tissues in solution are disrupted by gentle homogenization. This treatment ruptures the plasma membrane but leaves most of the organelles intact. The homogenate is then centrifuged; organelles such as nuclei, mitochondria, and lysosomes differ in size and therefore sediment at different rates. They also differ in specific gravity, and they "float" at different levels in a density gradient.

In figure 6 a tissue such as liver is first mechanically homogenized to break cells and disperse their contents in an aqueous buffer. The sucrose medium has an osmotic pressure similar to that in organelles, thus preventing diffusion of water into the organelles, which would swell and burst. (a) The large and small particles in the suspension can be separated by centrifugation at different speeds, or (b) particles of different density can be separated by isopycnic centrifugation. In isopycnic centrifugation, a centrifuge tube is filled with a solution, the density of which increases from top to bottom; a solute such as sucrose is dissolved at different concentrations to produce the density gradient. When a mixture of organelles is layered on top of the density gradient and the tube is centrifuged at high speed, individual organelles sediment until their buoyant density exactly matches that in the gradient. Each layer can be collected separately.

Differential centrifugation results in a rough fractionation of the cytoplasmic contents, which may be further purified by isopycnic ("same density") centrifugation. In this procedure, organelles of different buoyant densities (the result of different ratios of lipid and protein in each type of organelle) are separated on a density gradient. By carefully removing material from each region of the gradient and observing it with a microscope, the

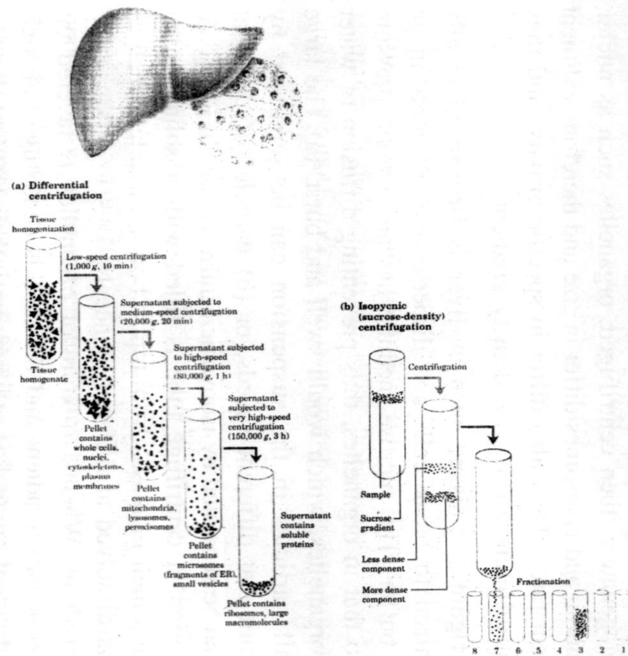

Figure 6: Subcellular fractionation of tissue

biochemist can establish the sedimentation position of each organelle and obtain purified organelles for further study. For example, these methods were used to establish that lysosomes contain degradative enzymes, mitochondria contain oxidative enzymes, and chloroplasts contain photosynthetic pigments. The isolation of an organelle enriched in a certain enzyme is often the first step in the purification of that enzyme.

Cytoplasm

Electron microscopy reveals several types of protein filaments crisscrossing the eukaryotic cell, forming an interlocking three-dimensional meshwork, the cytoskeleton. There are three general types of cytoplasmic filaments-actin filaments, microtubules, and intermediate filaments differing in width (from about 6 to 22 nm), composition, and specific function. All types provide structure and organization to the cytoplasm and shape to the cell. Actin filaments and microtubules also help to produce the motion of organelles or of the whole cell.

Each type of cytoskeletal component is composed of simple protein subunits that polymerize to form filaments of uniform thickness. These filaments are not permanent structures, they undergo constant disassembly into their protein subunits and reassembly into filaments. Their locations in cells are not rigidly fixed but may change dramatically with mitosis, cytokinesis, amoeboid motion, or changes in cell shape. The assembly, disassembly, and location of all types of filaments are regulated by other proteins, which serve to link or bundle the filaments or to move cytoplasmic organelles along the filaments.

The picture that emerges from this brief survey of cell structure is that of a eukaryotic cell with a meshwork of structural fibres and a complex system of membrane-bounded compartments. The filaments disassemble and then reassemble elsewhere. Membranous vesicles bud from one organelle and fuse with another. Organelles move through the cytoplasm along protein filaments, their motion powered by energy dependent motor proteins. The endomembrane system segregates specific metabolic processes and provides surfaces on which certain enzyme-catalyzed reactions occur. Exocytosis and endocytosis, mechanisms of transport (out of and into cells, respectively) that involve membrane fusion and fission, provide paths between the cytoplasm and surrounding medium, allowing for secretion of substances produced within the cell and uptake of extracellular materials.

Although complex, this organization of the cytoplasm is far from random. The motion and the positioning of organelles and cytoskeletal elements are under tight regulation, and at certain stages in a eukaryotic cell's life, dramatic, finely orchestrated reorganizations, such as the events of mitosis, occur. The interactions between the cytoskeleton and organelles are noncovalent, reversible, and subject to regulation in response to various intracellular and extracellular signals.

Supramolecular Structures

Macromolecules and their monomeric subunits differ greatly in size. A molecule of alanine is less than 0.5 nm long. Hemoglobin, the oxygen-carrying protein of erythrocytes (red blood cells), consists of nearly 600 amino acid subunits in four long chains, folded into globular shapes and associated in a structure 5.5 nm in diameter. In turn, proteins are much smaller than ribo-somes (about 20 nm in diameter), which are in turn much smaller than organelles such as mitochondria, typically 1,000 nm in diameter.

It is a long jump from simple biomolecules to cellular structures that can be seen with the light microscope. Figure 8 illustrates the structural hierarchy in cellular organization. The monomeric subunits in proteins, nucleic acids, and polysaccharides are joined by covalent bonds. In supramolecular complexes, however, macromolecules are held together by noncovalent interactions-much weaker, individually, than covalent bonds. Among these noncovalent interactions are

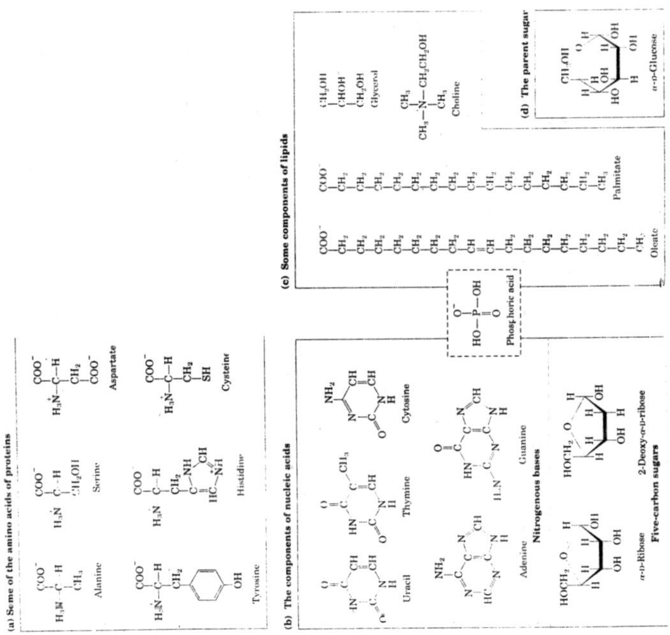

Figure 7: The organic compounds from which most cellular materials are constructed

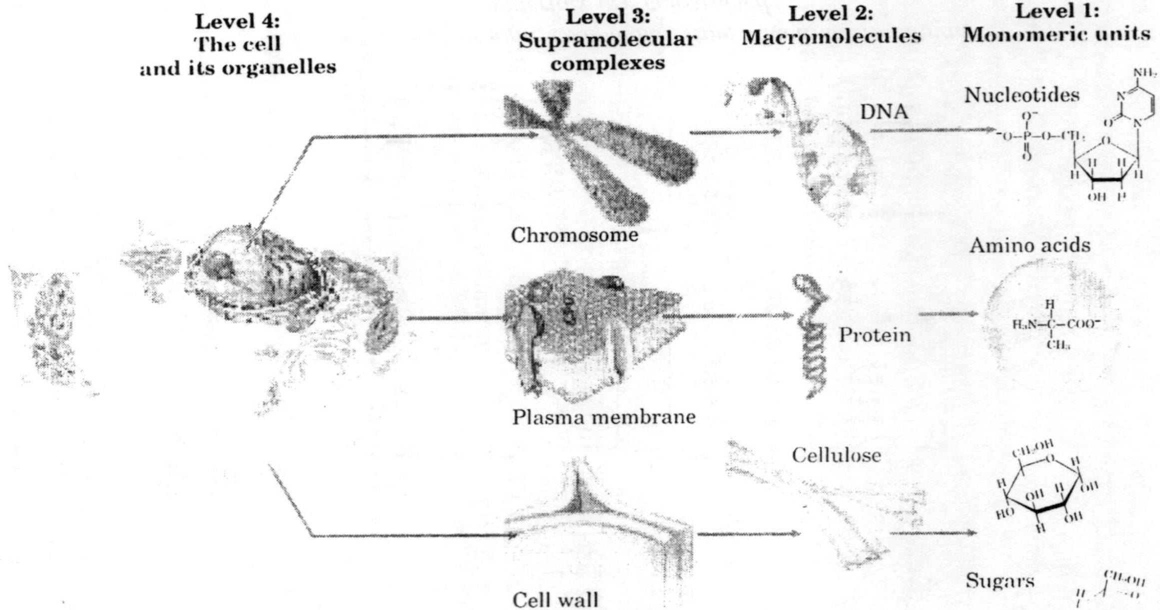

Figure 8: Structural hierarchy in the molecular organization of cells.

hydrogen bonds (between polar groups), ionic interactions (between charged groups), hydrophobic interactions (among nonpolar groups in aqueous solution), and van der Waals interactions-all of which have energies substantially smaller than those of covalent bonds (Table 1). The large numbers of weak interactions between macromolecules in supramolecular complexes stabilize these assemblies, producing their unique structures.

Table 1. Strengths of Bonds Common in Biomolecules

Type of bond	Bond dissociation energy (kJ/mol)	Type of bond	Bond dissociation energy (kJ/mol)
Single bonds		Double bonds	
O—H	470	C O	712
H H	435	C N	615
P O	419	C C	611
C H	414	P O	502
N H	389		
C O	352	Triple bonds	
C C	348	C C	816
S H	339	N N	930
C N	293		
C S	260		
N O	222		
S S	214		

Interactions among Molecules

One approach to understanding a biological process is to study purified molecules in vitro ("in glass"-in the test tube), without interference from other molecules present in the intact cell-that is, in vivo ("in the living"). Although this approach has been remarkably revealing, we must keep in mind that the inside of a cell is quite different from the inside of a test tube. The "interfering" components eliminated by purification may be critical to the biological function or regulation of the molecule purified. For example, in vitro studies of pure enzymes are commonly done at very low enzyme concentrations in thoroughly stirred aqueous solutions. In the cell, an enzyme is dissolved or suspended in a gel-like cytosol with thousands of other proteins, some of which bind to that enzyme and influence its activity.

Some enzymes are parts of multienzyme complexes in which reactants are channelled from one enzyme to another without ever entering the bulk solvent. Diffusion is hindered in the gel-like cytosol, and the cytosolic composition varies in different regions of the cell. In short, a given molecule may function quite differently in the cell than in vitro. A central challenge of biochemistry is to understand the influences of cellular organization and macromolecular associations on the function of individual enzymes and other biomolecules-to understand function in vivo as well as in vitro.

Chemical Foundations

Biochemistry aims to explain biological form and function in chemical terms. As we noted earlier, one of the most fruitful approaches to understanding biological phenomena has been to purify an individual chemical component, such as a protein, from a living organism and to characterize its structural and chemical characteristics. By the late eighteenth century, chemists had concluded that the composition of living matter is strikingly different from that of the inanimate world. Antoine Lavoisier noted the relative chemical simplicity of the "mineral world" and contrasted it with the complexity of the "plant and animal worlds"; the latter, he knew, were composed of compounds rich in the elements carbon, oxygen, nitrogen, and phosphorus.

During the first half of the twentieth century, parallel biochemical investigations of glucose breakdown in yeast and in animal muscle cells revealed remarkable chemical similarities in these two apparently very different cell types; the breakdown of glucose in yeast and muscle cells involved the same ten chemical intermediates. Subsequent studies of many other biochemical processes in many different organisms have confirmed the generality of this observation, neatly summarized by Jacques Monod: "What is true of *E*. coli is true of the elephant." The current understanding that all organisms share a common evolutionary origin is based in part on this observed universality of chemical intermediates and transformations.

Only about 30 of the more than 90 naturally occurring chemical elements are essential to organisms. Most of the elements in living matter have relatively low atomic numbers; only five have atomic numbers above that of selenium, 34. The four most abundant elements in living organisms, in terms of percentage of total number of atoms, are hydrogen, oxygen, nitrogen, and carbon, which together make up more than 99% of the mass of most cells.

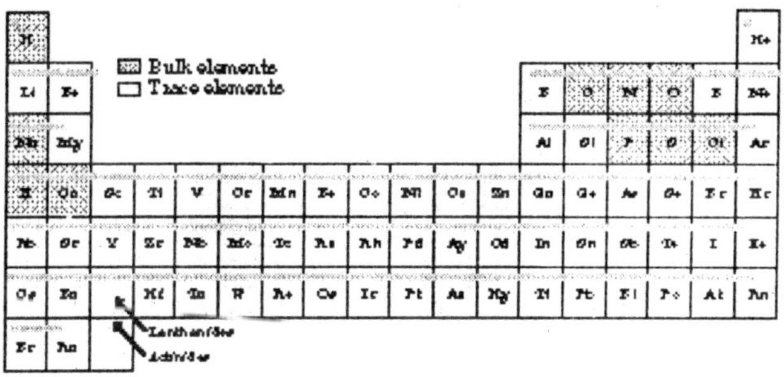

Figure 9: Elements essential to animal life and health

They are the lightest elements capable of forming one, two, three, and four bonds, respectively; in general, the lightest elements form the strongest bonds. The trace elements represent a miniscule fraction of the weight of the human body, but all are essential to life, usually because they are essential to the function of specific proteins, including enzymes. The oxygen-transporting capacity of the hemoglobin molecule, for example, is absolutely dependent on four iron ions that make up only 0.3% of its mass.

Biomolecules Are Compounds of Carbon

The chemistry of living organisms is organized around carbon, which accounts for more than half the dry weight of cells. Carbon can form single bonds with hydrogen atoms, and both single and double bonds with oxygen and nitrogen atoms. Of greatest significance in biology is the ability of carbon atoms to form very stable carbon-carbon single bonds. Each carbon atom can form single bonds with up to four other carbon atoms. Two carbon atoms also can share two (or three) electron pairs, thus forming double (or triple) bonds.

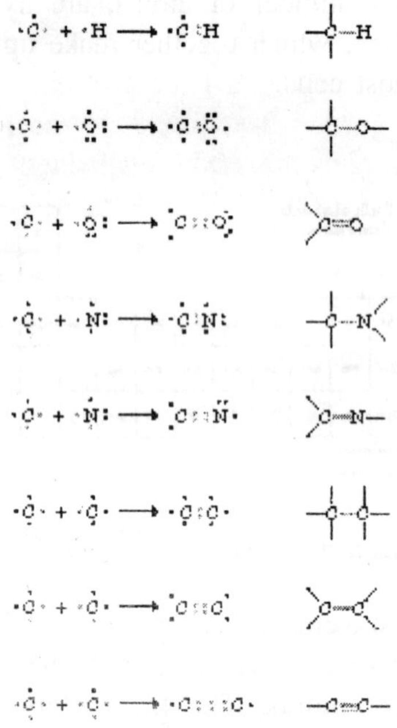

Figure 10: Versatility of carbon bonding.

The four single bonds that can be formed by a carbon atom are arranged tetrahedrally, with an angle of about 109.5° between any two bonds (Fig. 11) and an average length of 0.154 nm. There is free rotation around each single bond, unless very large or highly charged groups are attached to both carbon atoms, in which case rotation may be restricted. A double bond is shorter (about 0.134 nm) and rigid and allows little rotation about its axis.

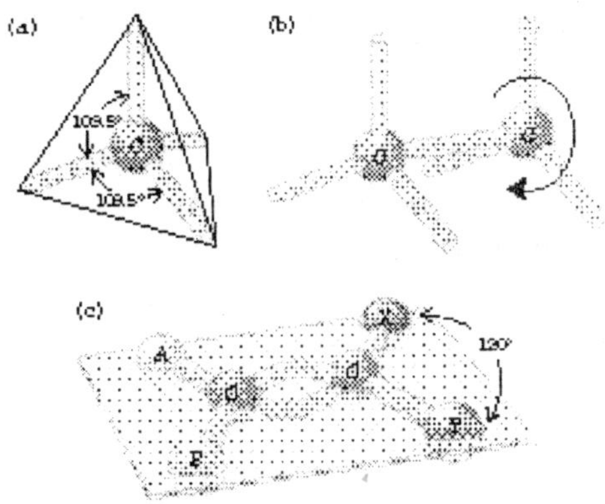

Figure 11: Geometry of carbon bonding.

In figure 11, (a) Carbon atoms have a characteristic tetrahedral arrangement of their four single bonds. (b) Carbon–carbon single bonds have freedom of rotation, as shown for the compound ethane (CH_3CH_3). (c) Double bonds are shorter and do not allow free rotation. The two doubly bonded carbons and the atoms designated A, B, X, and Y all lie in the same rigid plane.

Covalently linked carbon atoms in biomolecules can form linear chains, branched chains, and cyclic structures. To these carbon skeletons are added groups of other atoms, called functional groups, which confer specific chemical properties on the molecule. It seems likely that the bonding versatility of carbon was a major factor in the selection of carbon compounds for the molecular machinery of cells during the origin and evolution of living organisms. No other chemical element can form molecules of such widely different sizes and shapes or with such a variety of functional groups.

Most biomolecules can be regarded as derivatives of hydrocarbons, with hydrogen atoms replaced by a variety of functional groups to yield different families of organic compounds. Typical of these are alcohols, which have one or more hydroxyl groups; amines, with amino groups; aldehydes and ketones, with carbonyl groups; and carboxylic acids, with carboxyl groups (Fig. 12). Many biomolecules are polyfunctional, containing two or more different kinds of functional groups (Fig. 13), each with its own chemical characteristics and reactions. The chemical "personality" of a compound is determined by the chemistry of its functional groups and their disposition in three-dimensional space.

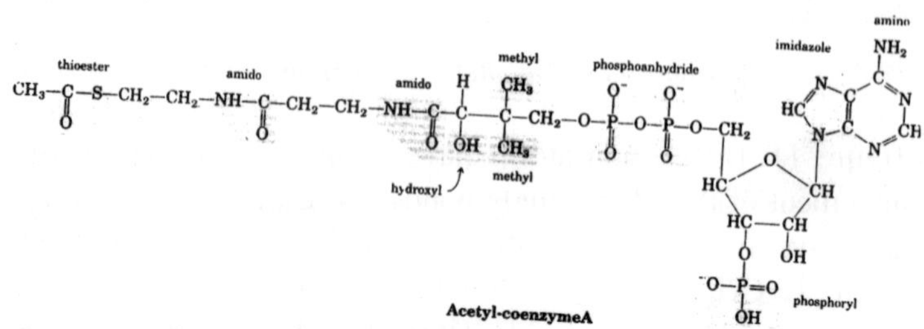

Figure 13: Several common functional groups in a single biomolecule

Universal Set of Small Molecules

Dissolved in the aqueous phase (cytosol) of all cells is a collection of 100 to 200 different small organic molecules (M_r

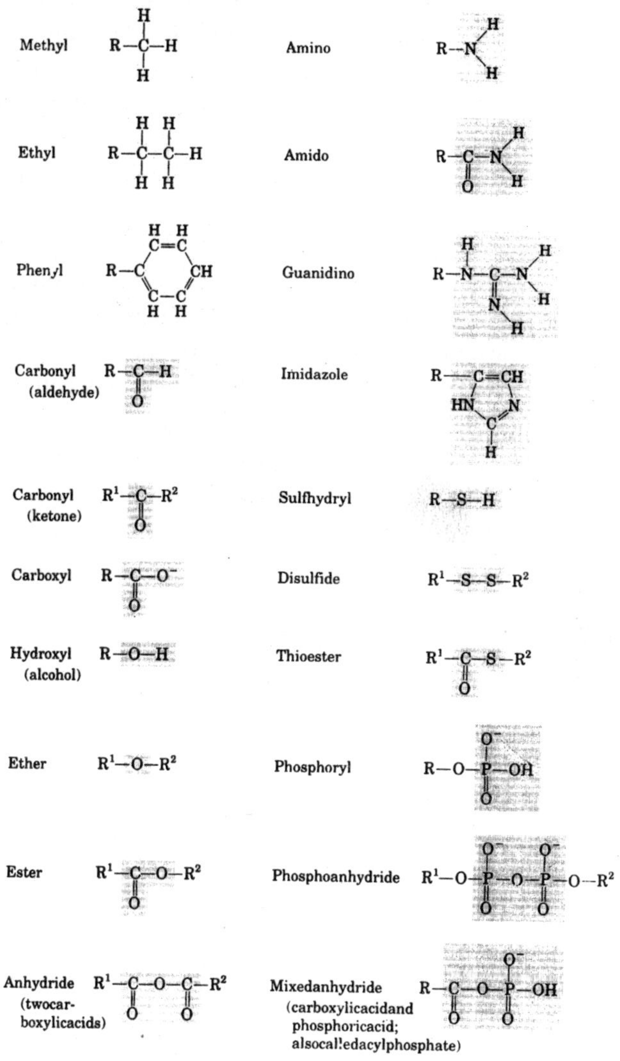

Figure 12: Some common functional groups of biomolecules

~100 to ~500), the central metabolites in the major pathways occurring in nearly every cell-the metabolites and pathways that have been conserved throughout the course of evolution. This collection of molecules includes the common amino acids, nucleotides, sugars and their phosphorylated derivatives, and a number of mono-, di-, and tricarboxylic acids. The molecules are polar or charged, water soluble, and present in micromolar to millimolar concentrations. They are trapped within the cell because the plasma membrane is impermeable to them-although specific membrane transporters can catalyze the movement of some molecules into and out of the cell or between compartments in eukaryotic cells. The universal occurrence of the same set of compounds in living cells is a manifestation of the universality of metabolic design, reflecting the evolutionary conservation of metabolic pathways that developed in the earliest cells.

There are other small biomolecules, specific to certain types of cells or organisms. For example, vascular plants contain, in addition to the universal set, small molecules called secondary metabolites, which play a role specific to plant life. These metabolites include compounds that give plants their characteristic scents, and compounds such as morphine, quinine, nicotine, and caffeine that are valued for their physiological effects on humans but used for other purposes by plants. The entire collection of small molecules in a given cell has been called that cell's metabolome, in parallel with the term "genome". If we knew the composition of a cell's metabolome, we could predict which enzymes and metabolic pathways were active in that cell.

Many biological molecules are macromolecules, polymers of high molecular weight assembled from relatively simple precursors. Proteins, nucleic acids, and polysaccharides are produced by the polymerization of relatively small compounds with molecular weights of 500 or less. The number of polymerized units can range from tens to millions. Synthesis of macromolecules is a major energy-consuming activity of cells.

Macromolecules themselves may be further assembled into supramolecular complexes, forming functional units such as ribosomes. Table 2 shows the major classes of biomolecules in the bacterium *E.coli*.

Table 2. Molecular Components of an E. coli Cell

	Percentage of total weight of cell	Approximate number of different molecular species
Water	70	1
Proteins	15	3,000
Nucleic acids		
DNA	1	1
RNA	6	>3,000
Polysaccharides	3	5
Lipids	2	20
Monomeric subunits and intermediates	2	500
Inorganic ions	1	20

Proteins, long polymers of amino acids, constitute the largest fraction (besides water) of cells. Some proteins have catalytic activity and function as enzymes; others serve as structural elements, signal receptors, or transporters that carry specific substances into or out of cells. Proteins are perhaps the most versatile of all biomolecules. The nucleic acids, DNA and RNA, are polymers of nucleotides. They store and transmit genetic information, and some RNA molecules have structural and catalytic roles in supramolecular complexes. The polysaccharides, polymers of simple sugars such as glucose, have two major functions: as energy-yielding fuel stores and as extracellular structural elements with specific binding sites for particular proteins. Shorter polymers of sugars (oligosaccharides) attached to proteins or lipids at the cell surface serve as specific cellular signals. The lipids, greasy or oily hydrocarbon derivatives, serve as structural components of membranes, energy-rich fuel stores, pigments, and intracellular signals. In proteins, nucleotides, polysaccharides, and lipids, the number of monomeric subunits is very large: molecular weights in the range of 5,000 to more than 1 million for proteins, up to several billion for nucleic acids, and in the millions for polysaccharides such as starch. Individual lipid molecules are much smaller (M_r 750 to 1,500) and are not classified as macromolecules. However, large numbers of lipid molecules can associate noncovalently into very large structures. Cellular membranes are built of enormous noncovalent aggregates of lipid and protein molecules.

Proteins and nucleic acids are informational macromolecules: each protein and each nucleic acid has a characteristic information-rich subunit sequence. Some oligosaccharides, with six or more different sugars connected in branched chains, also carry information; on the outer surface of cells they serve as highly specific points of recognition in many cellular processes.

Three-Dimensional Structure

The covalent bonds and functional groups of a biomolecule are, of course, central to its function, but so also is the arrangement of the molecule's constituent atoms in three-dimensional space-its stereochemistry. A carbon-containing compound commonly exists as stereoisomers, molecules with the same chemical bonds but different stereochemistry-that is, different configuration, the fixed spatial arrangement of atoms. Interactions between biomolecules are invariably stereo-specific, requiring specific stereochemistry in the interacting molecules.

Figure 14 shows three ways to illustrate the stereo-chemical structures of simple molecules. The perspective diagram specifies stereochemistry unambiguously, but bond angles and centre-to-centre bond lengths are better represented with ball-and-stick models. In space-filling models, the radius of each atom is proportional to its van der Waals radius, and the contours of the model define the space occupied by the molecule (the volume of space from which atoms of other molecules are excluded).

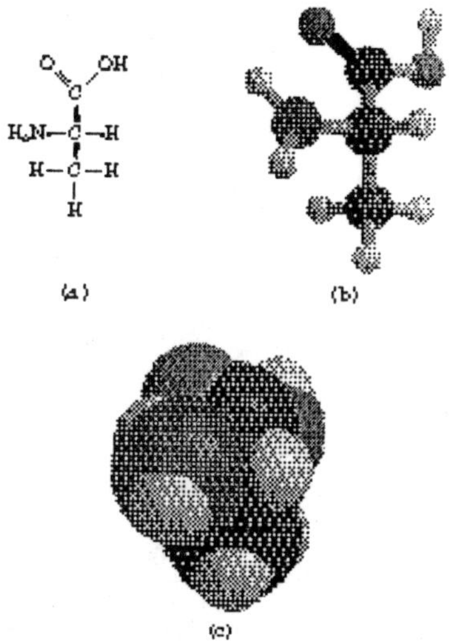

Figure 14: Representations of molecules.

Configuration is conferred by the presence of either (1) double bonds, around which there is no freedom of rotation, or (2) chiral centers, around which substituent groups are arranged in a specific sequence. The identifying characteristic of configurational isomers is that they cannot be interconverted without temporarily breaking one or more covalent bonds. Figure 15a shows the configurations of maleic acid and its isomer, fumaric acid. These compounds are geometric, or cistrans, isomers; they differ in the arrangement of their substituent groups with respect to the nonrotating double bond (Latin cis, "on this side"-groups on the same side of the double bond; trans, "across"-groups on opposite sides). Maleic acid is the cis isomer and fumaric acid the trans isomer; each is a well-defined

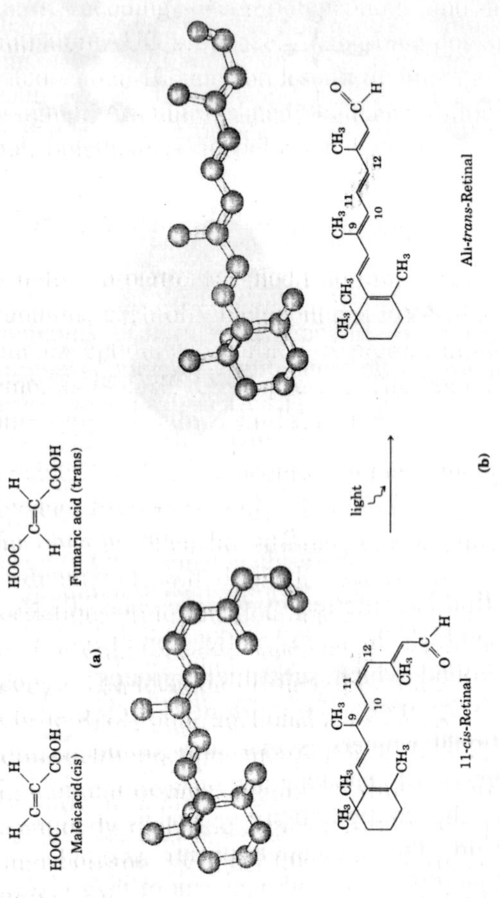

Figure 15: Configurations of geometric isomers.

compound that can be separated from the other, and each has its own unique chemical properties. A binding site that is complementary to one of these molecules would not be a suitable binding site for the other, which explains why the two compounds have distinct biological roles despite their similar chemistry.

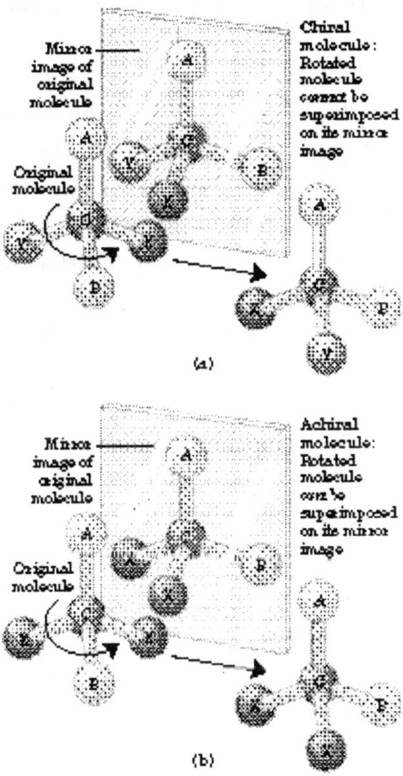

Figure 16: Molecular asymmetry: chiral and achiral molecules.

In the second type of configurational isomer, four different substituents bonded to a tetrahedral carbon atom may be arranged two different ways in space-that is, have two configurations (Fig. 16)-yielding two stereoisomers with similar or identical chemical properties but differing in certain physical and biological properties. A carbon atom with four different substituents is said to be asymmetric, and asymmetric carbons are called chiral centers (Greek chiros, "hand"; some stereoisomers are related structurally as the right hand is to the left). A molecule with only one chiral carbon can have two stereoisomers; when two or more (n) chiral carbons are present, there can be 2^n stereoisomers. Some stereoisomers are mirror images of each other; they are called enantiomers (Fig. 16). Pairs of stereoisomers that are not mirror images of each other are called diastereomers (Fig. 17).

As Louis Pasteur first observed, enantiomers have nearly identical chemical properties but differ in a characteristic physical property, their interaction with plane-polarized light. In separate solutions, two enantiomers rotate the plane of plane-polarized light in opposite directions, but an equimolar solution of the two enantiomers (a racemic mixture) shows no optical rotation. Compounds without chiral centers do not rotate the plane of plane-polarized light.

Given the importance of stereochemistry in reactions between biomolecules, biochemists must name and represent the structure of each bio-molecule so that its stereochemistry is unambiguous. For compounds with more than one chiral centre, the most useful system of nomenclature is the RS system. In this system, each group attached to a chiral carbon is assigned a priority. The priorities of some common substituents are

$$-OCH_2 >- OH >- NH_2 >- COOH >- CHO >- CH_2OH >- CH_3 >-H$$

For naming in the RS system, the chiral atom is viewed with the group of lowest priority (4 in the diagram on the next page)

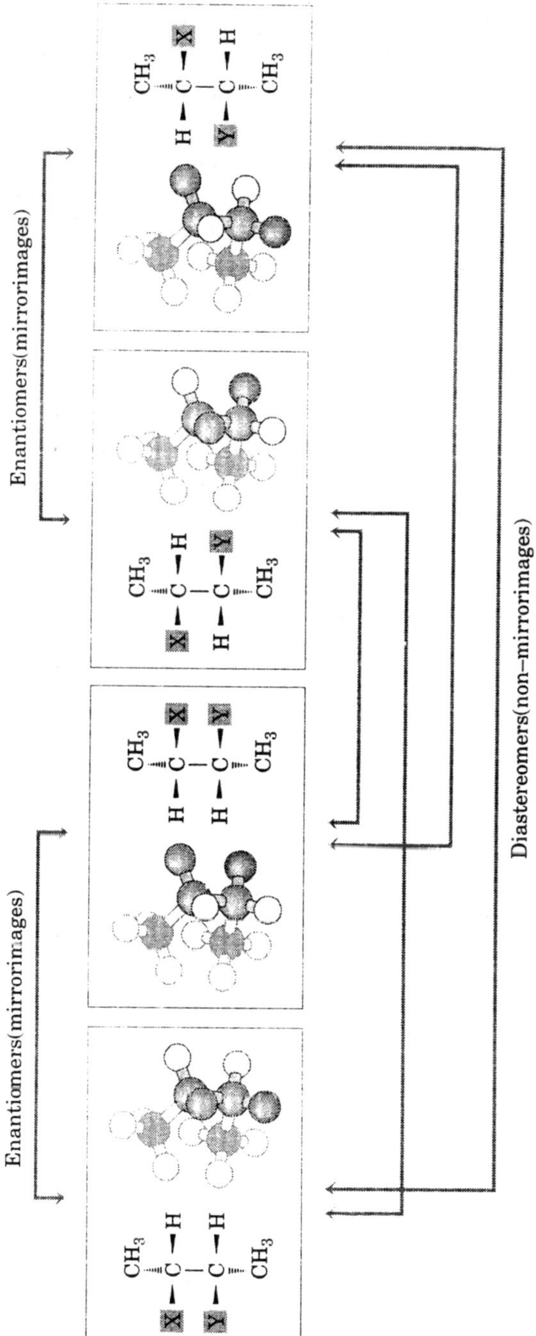

Figure 17: Two types of stereoisomers.

pointing away from the viewer. If the pri-ority of the other three groups (1 to 3) decreases in clockwise order, the configuration is *(R)* (Latin rectus, "right"); if in counterclockwise order, the configuration is (S) (Latin sinister, "left"). In this way each chiral carbon is designated either *(R)* or (S), and the inclusion of these designations in the name of the compound provides an unambiguous description of the stereochemistry at each chiral centre.

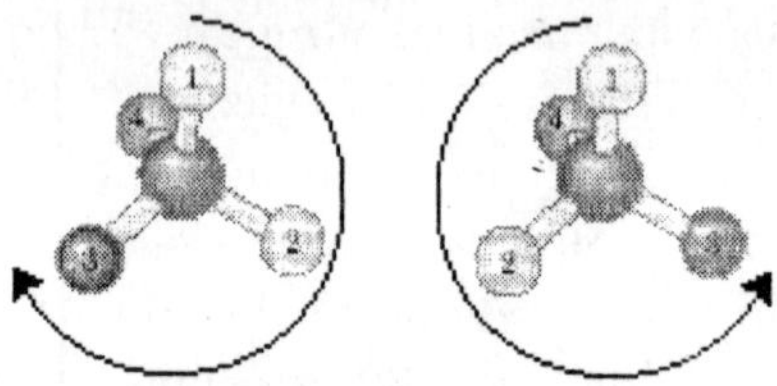

Clockwise (*R*) Counterclockwise (*S*)

A molecule with a single chiral centre (glyceraldehydes, for example) can be named unambiguously by either system.

L-Glyceraldehyde (*S*)-Glyceraldehyde

Distinct from configuration is molecular conformation, the spatial arrangement of substituent groups that, without breaking any bonds, are free to assume different positions in space because of the freedom of rotation about single bonds. In the simple hydrocarbon ethane, for example, there is nearly complete freedom of rotation around the C C bond. Many different, interconvertible conformations of ethane are possible, depending on the degree of rotation (Fig. 18).

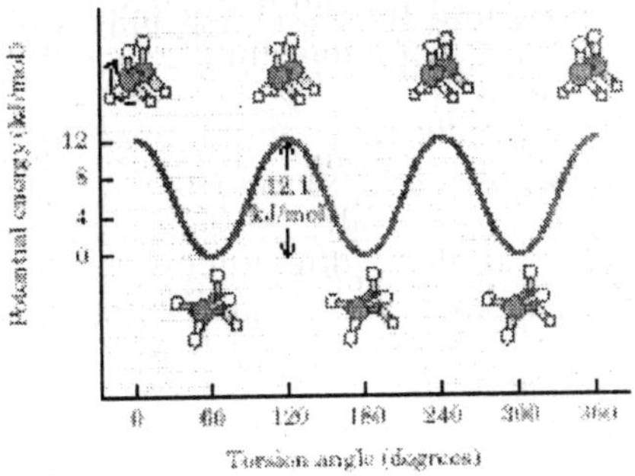

Figure 18: Conformations

Two conformations are of special interest: the staggered, which is more stable than all others and thus predominates, and the eclipsed, which is least stable. We cannot isolate either of these conformational forms, because they are freely interconvertible. However, when one or more of the hydrogen atoms on each carbon is replaced by a functional group that is either very large or electrically charged, freedom of rotation around the C C bond is hindered. This limits the number of stable conformations of the ethane derivative.

Interactions between Biomolecules Are Stereospecific

Biological interactions between molecules are stereo-specific: the "fit" in such interactions must be stereo-chemically correct. The three-dimensional structure of biomolecules large and small-the combination of configuration and conformation-is of the utmost importance in their biological interactions: reactant with enzyme, hormone with its receptor on a cell surface, antigen with its specific antibody, for example. The study of biomolecular stereochemistry with precise physical methods is an important part of modern research on cell structure and biochemical function.

In living organisms, chiral molecules are usually present in only one of their chiral forms. For example, the amino acids in proteins occur only as their L isomers; glucose occurs only as its D isomer. In contrast, when a compound with an asymmetric carbon atom is chemically synthesized in the laboratory, the reaction usually produces all possible chiral forms: a mixture of the D and L forms, for example. Living cells produce only one chiral form of biomolecules because the enzymes that synthesize them are also chiral.

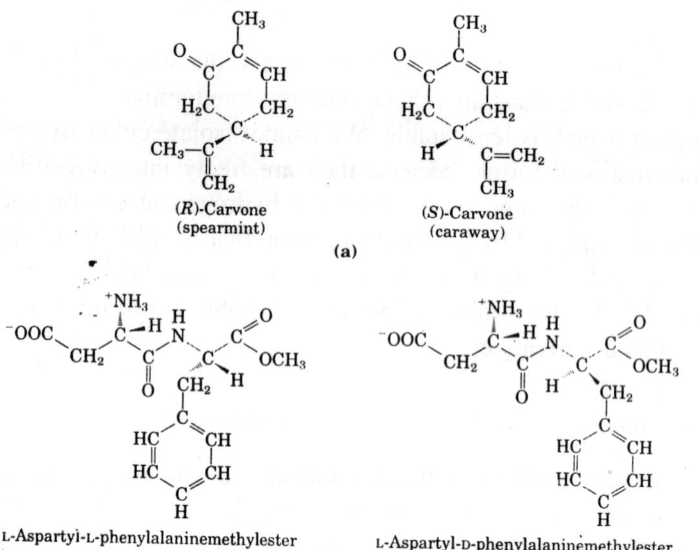

(*R*)-Carvone (spearmint)

(*S*)-Carvone (caraway)

(a)

L-Aspartyl-L-phenylalaninemethylester (aspartame)(sweet)

L-Aspartyl-D-phenylalaninemethylester (bitter)

(b)

Figure 19: Stereoisomers distinguishable by smell and taste in humans

Stereospecificity, the ability to distinguish between stereoisomers, is a property of enzymes and other proteins and a characteristic feature of the molecular logic of living cells. If the binding site on a protein is complementary to one isomer of a chiral compound, it will not be complementary to the other isomer, for the same reason that a left glove does not fit a right hand. Two striking examples of the ability of biological systems to distinguish stereoisomers are shown in Figure 19.

Physical Foundations

Living cells and organisms must perform work to stay alive and to reproduce themselves. The synthetic reactions that occur within cells, like the synthetic processes in any factory, require the input of energy. Energy is also consumed in the motion of a bacterium or an Olympic sprinter, in the flashing of a firefly or the electrical discharge of an eel. And the storage and expression of information require energy, without which structures rich in information inevitably become disordered and meaningless.

In the course of evolution, cells have developed highly efficient mechanisms for coupling the energy obtained from sunlight or fuels to the many energy-consuming processes they must carry out. One goal of biochemistry is to understand, in quantitative and chemical terms, the means by which energy is extracted, channelled, and consumed in living cells. We can consider cellular energy conversions-like all other energy conversions-in the context of the laws of thermodynamics.

Living Organisms in a Dynamic Steady State

The molecules and ions contained within a living organism differ in kind and in concentration from those in the organism's surroundings. A Paramecium in a pond, a shark in the ocean, an erythrocyte in the human bloodstream, an apple tree in an orchard-all are different in composition from their surroundings and, once they have reached maturity, all maintain a constant composition in the face of constantly changing surroundings.

Although the characteristic composition of an organism changes little through time, the population of molecules within the organism is far from static. Small molecules, macromolecules, and supramolecular complexes are continuously synthesized and then broken down in chemical reactions that involve a constant flux of mass and energy through the system. The hemoglo-bin molecules carrying oxygen from your lungs to your brain at this moment were synthesized within the past month; by next month they will have been degraded and entirely replaced by new hemoglobin molecules. The glucose you ingested with your most recent meal is now circulating in your bloodstream; before the day is over these particular glucose molecules will have been converted into something else-carbon dioxide or fat, perhaps-and will have been replaced with a fresh supply of glucose, so that your blood glucose concentration is more or less constant over the whole day.

The amounts of hemoglobin and glucose in the blood remain nearly constant because the rate of synthesis or intake of each just balances the rate of its breakdown, consumption, or conversion into some other product. The constancy of concentration is the result of a dynamic steady state, a steady state that is far from equilibrium. Maintaining this steady state requires the constant investment of energy; when the cell can no longer generate energy, it dies and begins to decay toward equilibrium with its surroundings. We consider below exactly what is meant by "steady state" and "equilibrium."

Organisms Transform Energy

For chemical reactions occurring in solution, we can define a system as all the reactants and products present, the solvent that contains them, and the immediate at-mosphere-in short, everything within a defined region of space. The system and its surroundings together constitute the universe. If the system exchanges neither matter nor energy with its surroundings, it is said to be isolated. If the system exchanges energy but not matter with its surroundings, it is a closed system; if it exchanges both energy and matter with its surroundings, it is an open system.

A living organism is an open system; it exchanges both matter and energy with its surroundings. Living organisms derive energy from their surroundings in two ways: (1) they take up chemical fuels from the environment and extract energy by oxidizing them; or (2) they absorb energy from sunlight.

The first law of thermodynamics, developed from physics and chemistry but fully valid for biological systems as well, describes the principle of the conservation of energy: in any physical or chemical change, the total amount of energy in the universe remains constant, although the form of the energy may change. Cells are consummate transducers of energy, capable of interconverting chemical, electromagnetic, mechanical, and osmotic energy with great efficiency.

Flow of Electrons

Nearly all living organisms derive their energy, directly or indirectly, from the radiant energy of sunlight, which arises from thermonuclear fusion reactions carried out in the sun. Photosynthetic cells absorb light energy and use it to drive electrons from water to carbon dioxide, forming energy-rich products such as glucose ($C_6 H_{12} O_6$), starch, and sucrose and releasing O_2 into the atmosphere:

$$6CO_2\ 6H_2O \xrightarrow{\text{light}} C_6H1_2O_6\ 6O_2$$

(light-driven reduction of CO_2)

Nonphotosynthetic cells and organisms obtain the en-ergy they need by oxidizing the energy-rich products of photosynthesis and then passing electrons to atmospheric O_2 to form water, carbon dioxide, and other end products, which are recycled in the environment:

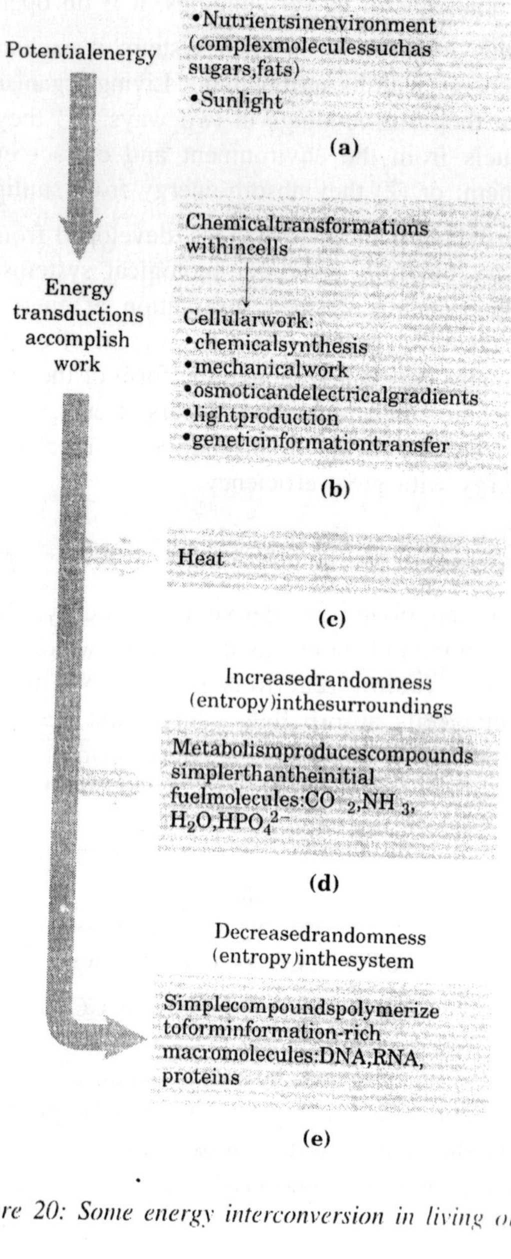

Figure 20: Some energy interconversion in living organisms

$$C_6H_{12}O_6 + O_2 \; 6CO_2 + 6H_2O + \text{energy}$$

(energy-yielding oxidation of glucose)

Virtually all energy transductions in cells can be traced to this flow of electrons from one molecule to another, in a "downhill" flow from higher to lower electrochemical potential; as such, this is formally analogous to the flow of electrons in a battery-driven electric circuit. All these reactions involving electron flow are oxidation-reduction reactions: one reactant is oxidized (loses electrons) as another is reduced (gains electrons).

Creating and Maintaining Order Requires Work and Energy

DNA, RNA, and proteins are informational macromolecules. In addition to using chemical energy to form the covalent bonds between the subunits in these polymers, the cell must invest energy to order the subunits in their correct sequence. It is extremely improbable that amino acids in a mixture would spontaneously condense into a single type of protein, with a unique sequence. This would represent increased order in a population of molecules; but according to the second law of thermodynamics, the tendency in nature is toward ever-greater disorder in the universe: the total entropy of the universe is continually increasing. To bring about the synthesis of macromolecules from their monomeric units, free energy must be supplied to the system (in this case, the cell).

The randomness or disorder of the components of a chemical system is expressed as entropy, S . Any change in randomness of the system is expressed as entropy change, ΔS, which by convention has a positive value when randomness increases. J. Willard Gibbs, who developed the theory of energy changes during chemical reactions, showed that the free-energy content, G, of any closed system can be defined in terms of three quantities: enthalpy, H, reflecting the number and kinds of bonds; entropy, S; and the absolute temperature, T (in degrees Kelvin). The definition of free energy is $G = H - TS$. When a chemical reaction occurs at constant temperature, the free-energy change, ΔG, is determined by the enthalpy change, ΔH, reflecting the kinds and numbers of chemical bonds and noncovalent interactions broken and formed, and the entropy change, ΔS, describing the change in the system's randomness:

$$\Delta G = \Delta H - T\,\Delta S$$

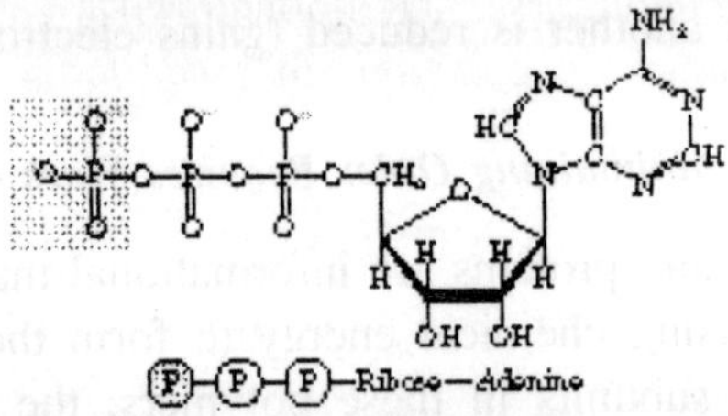

Figure 21: Adenosine triphosphate (ATP)

A process tends to occur spontaneously only if G is negative. Yet cell function depends largely on molecules, such as proteins and nucleic acids, for which the free energy of formation is positive: the molecules are less stable and more highly ordered than a mixture of their monomeric components. To carry out these thermodynamically unfavourable, energy-requiring (endergonic) reactions, cells couple them to other reactions that liberate free energy (exergonic reactions), so that the overall process is exergonic: the sum of the free-energy changes is negative. The usual source of free energy in coupled biological reactions is the energy released by hydrolysis of phosphoanhydride bonds such as those in adenosine triphosphate. In figure 21 each P represents a phosphoryl group:

Amino acids polymer ΔG_1 is positive (endergonic)

—(P)—(P) ⟶ —(P) + (P) ΔG_2 is negative (exergonic)

When these reactions are coupled, the sum of ΔG_1 and ΔG_2 is negative-the overall process is exergonic. By this coupling strategy, cells are able to synthesize and maintain the information-rich polymers essential to life.

Energy Coupling Links Reactions

The central issue in bioenergetics is the means by which energy from fuel metabolism or light capture is coupled to a cell's energy-requiring reactions. In thinking about energy coupling, it is useful to consider a simple mechanical example, as shown in Figure 22a. An object at the top of an inclined plane has a certain amount of potential energy as a result of its elevation. It tends to slide down the plane, losing its potential energy of position as it approaches the ground. When an appropriate string-and-pulley device couples the falling object to another, smaller object, the spontaneous downward motion of the larger can lift the smaller, accomplishing a certain amount of work. The amount of energy available to do work is the free-energy change, ΔG; this is always somewhat less than the theoretical amount of energy released, because some energy is dissipated as the heat of friction. The greater the elevation of the larger object, the greater the energy released (ΔG) as the object slides downward and the greater the amount of work that can be accomplished.

How does this apply in chemical reactions? In closed systems, chemical reactions proceed spontaneously until equilibrium is reached. When a system is at equilibrium, the rate of product formation exactly equals the rate at which product is converted to reactant. Thus there is no net change in the concentration of reactants and products; a steady state is achieved. The energy change as the system moves from its initial state to equilibrium, with no changes in temperature or pressure, is given by the free-energy change, ΔG. The magnitude of ΔG depends on the particular chemical reaction and on how far from equilibrium the system is initially. Each compound involved in a chemical reaction contains a certain amount of potential energy, related to the kind and number of its bonds. In reactions that occur spontaneously, the products have less free energy than the reactants, thus the reaction releases free energy, which is then available to do work. Such reactions are exergonic; the decline in free energy from reactants to products is expressed as a negative value. Endergonic reactions require an input of energy, and their ΔG values are positive. As in mechanical processes, only part of the energy released in exergonic chemical reactions can be used to accomplish work. In living systems some energy is dissipated as heat or lost to increasing entropy.

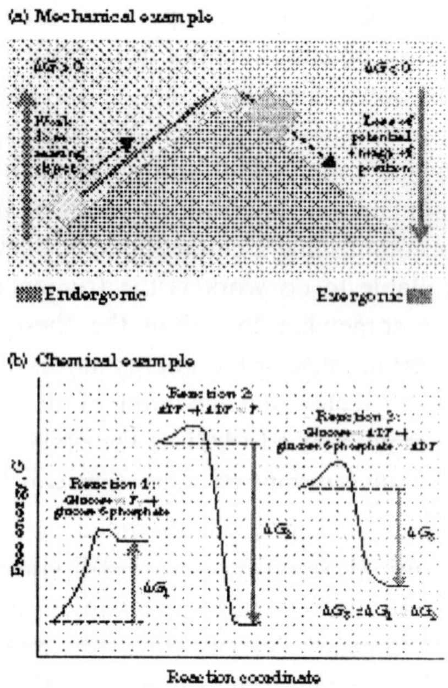

Figure 22: Energy coupling in mechanical and chemical processes.

In living organisms, as in the mechanical example in Figure 22a, an exergonic reaction can be coupled to an endergonic reaction to drive otherwise unfavourable reactions. Figure 22b (a type of graph called a reaction coordinate diagram) illustrates this principle for the conversion of glucose to glucose 6-phosphate, the first step in the pathway for oxidation of glucose. The simplest way to produce glucose 6-phosphate would be:

Reaction 1: Glucose + P_i glucose 6-phosphate
(endergonic; ΔG_1 is positive)

This reaction does not occur spontaneously; ΔG is positive. A second, very exergonic reaction can occur in all cells:

Reaction 2: ATP ADP P_i
(exergonic; ΔG_2 is negative)

These two chemical reactions share a common intermediate, P_i, which is consumed in reaction 1 and produced in reaction 2. The two reactions can therefore be coupled in the form of a third reaction, which we can write as the sum of reactions 1 and 2, with the common intermediate, P_i, omitted from both sides of the equation:

Reaction 3: Glucose + ATP glucose 6-phosphate + ADP

Because more energy is released in reaction 2 than is consumed in reaction 1, the free-energy change for reaction 3, ΔG_3, is negative, and the synthesis of glucose 6-phosphate can therefore occur by reaction 3. The coupling of exergonic and endergonic reactions through a shared intermediate is absolutely central to the energy exchanges in living systems. As we shall see, the breakdown of ATP (reaction 2 in Fig. 22b) is the ex-ergonic reaction that drives many endergonic processes in cells. In fact, ATP is the major carrier of chemical energy in all cells.

K_{eq} *and* $\Delta G°$ *Measures*

The tendency of a chemical reaction to go to completion can be expressed as an equilibrium constant. For the reaction aA + bB cC + dD

the equilibrium constant, K_{eq}, is given by

$$K_{eq} = \frac{[C_{eq}]^c[D_{eq}]^d}{[A_{eq}]^a[B_{eq}]^b}$$

where $[A_{eq}]$ is the concentration of A, $[B_{eq}]$ the concen-tration of B, and so on, when the system has reached equilibrium. A large value of K_{eq} meańs the reaction tends to proceed until the reactants have been almost completely converted into the products.

Gibbs showed that ΔG for any chemical reaction is a function of the standard free-energy change, $\Delta G°$- a constant that is characteristic of each specific reaction-and a term that expresses the initial concentrations of reactants and products:

$$\Delta G = \Delta G° + RT \ln \frac{[C_i]^c[D_i]^d}{[A_i]^a[B_i]^b} \qquad (1)$$

where $[A_i]$ is the initial concentration of A, and so forth; R is the gas constant; and T is the absolute temperature. When a reaction has reached equilibrium, no driving force remains and it can do no work: $\Delta G = 0$. For this special case, $[A_i]$ $[A_{eq}]$, and so on, for all reactants and products, and

$$\frac{[C_i]^c[D_i]^d}{[A_i]^a[B_i]^b} = \frac{[C_{eq}]^c[D_{eq}]^d}{[A_{eq}]^a[B_{eq}]^b} = K_{eq}$$

Substituting 0 for ΔG and K_{eq} for $[C_i]^c[D_i]^d/[A_i]^a[B_i]^b$ in Equation 1-1, we obtain the relationship

$$\Delta G° = - RT \ln K_{eq}$$

from which we see that ΔG° is simply a second way (besides K_{eq}) of expressing the driving force on a reaction. Because K_{eq} is experimentally measurable, we have a way of determining ΔG°, the thermodynamic constant characteristic of each reaction.

The units of ΔG° and ΔG are joules per mole (or calories per mole). When $K_{eq} >> 1$, ΔG° is large and negative; when $K_{eq} << 1$, ΔG° is large and positive. From a table of experimentally determined values of either K_{eq} or ΔG°, we can see at a glance which reactions tend to go to completion and which do not.

One caution about the interpretation of ΔG°: thermodynamic constants such as this show where the final equilibrium for a reaction lies but tell us nothing about how fast that equilibrium will be achieved.

Enzymes Promote Sequences

All biological macromolecules are much less thermodynamically stable than their monomeric subunits, yet they are kinetically stable: their uncatalyzed break-down occurs so slowly (over years rather than seconds) that, on a time scale that matters for the organism, these molecules are stable. Virtually every chemical reaction in a cell occurs at a significant rate only because of the presence of enzymes-biocatalysts that, like all other catalysts, greatly enhance the rate of specific chemical reactions without being consumed in the process.

The path from reactant(s) to product(s) almost invariably involves an energy barrier, called the activation barrier (Fig. 23), that must be surmounted for any reaction to proceed. The breaking of existing bonds and formation of new ones generally requires, first, the distortion of the existing bonds, creating a transition state of higher free energy than either reactant or product. The highest point in the reaction coordinate diagram represents the transition state, and the difference in energy between the reactant in its ground state and in its transition state is the activation energy, $\Delta G^\ddagger$. An enzyme catalyzes a reaction by providing a more comfortable fit for the transition state: a surface that complements the transition state in stereochemistry, plurality, and charge. The binding of enzyme to the transition state is exergonic, and the energy released by this binding reduces the activation energy for the reaction and greatly increases the reaction rate.

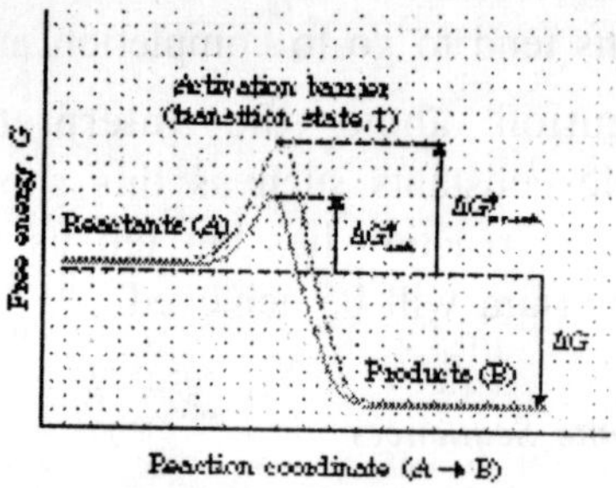

Figure 23: Energy changes during a chemical reaction

A further contribution to catalysis occurs when two or more reactants bind to the enzyme's surface close to each other and with stereospecific orientations that favour the reaction. This increases by orders of magnitude the probability of productive collisions between reactants.

Cellular catalysts are, with a few exceptions, proteins. Again with a few exceptions, each enzyme catalyzes a specific reaction, and each reaction in a cell is catalyzed by a different enzyme. Thousands of different enzymes are therefore required by each cell. The multiplicity of enzymes, their specificity (the ability to discriminate between reactants), and their susceptibility to regulation give cells the capacity to lower activation barriers selectively. This selectivity is crucial for the effective regulation of cellular processes. By allowing specific reactions to proceed at significant rates at particular times, enzymes determine how matter and energy are channeled into cellular activities.

The thousands of enzyme-catalyzed chemical reactions in cells are functionally organized into many sequences of consecutive reactions, called pathways, in which the product of one reaction becomes the reactant in the next. Some pathways degrade organic nutrients into simple end products in order to extract chemical energy and convert it into a form useful to the cell; together these degradative, free-energy-yielding reactions are designated catabolism. Other pathways start with small precursor molecules and convert them to progressively larger and more complex molecules, including proteins and nucleic acids. Such synthetic pathways, which invariably require the input of energy, are collectively designated anabolism. The overall network of enzyme-catalyzed pathways constitutes cellular metabolism. ATP is the major connecting link (the shared intermediate) between the catabolic and anabolic components of this network (shown schematically in Fig. 24). The pathways of enzyme-catalyzed reactions that act on the main constituents of cells-proteins, fats, sugars, and nucleic acids-are virtually identical in all living organisms.

To Achieve Balance and Economy

Not only do living cells simultaneously synthesize thousands of different kinds of carbohydrate, fat, protein, and nucleic acid molecules and their simpler subunits, but they do so in the precise proportions required by the cell under any given circumstance. For example, during rapid cell growth the precursors of proteins and nucleic acids must be made in large quantities, whereas in nongrowing cells the requirement for these precursors is much reduced. Key enzymes in each metabolic pathway are regulated so that each type of precursor molecule is produced in a quantity appropriate to the current requirements of the cell.

Consider the pathway in *E.* coli that leads to the synthesis of the amino acid isoleucine, a constituent of proteins. The pathway has five steps catalyzed by five different enzymes (A through F represent the intermediates in the pathway):

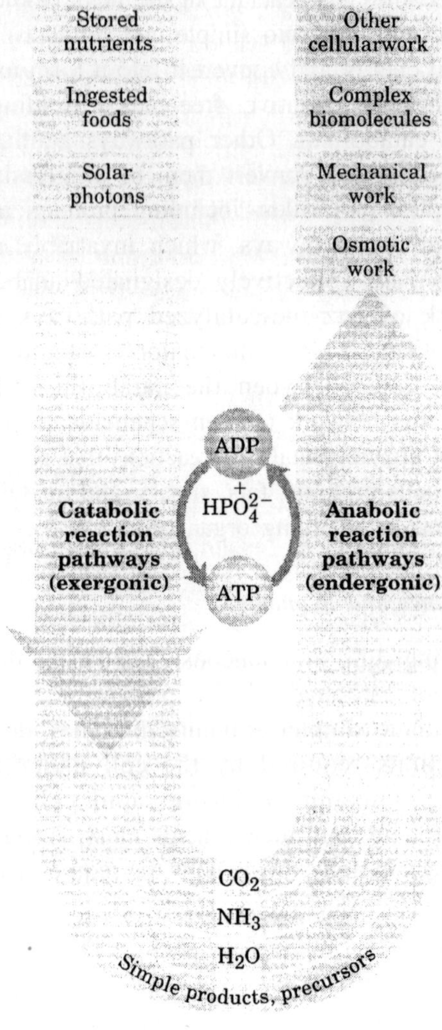

Figure 24: The central role of ATP in metabolism

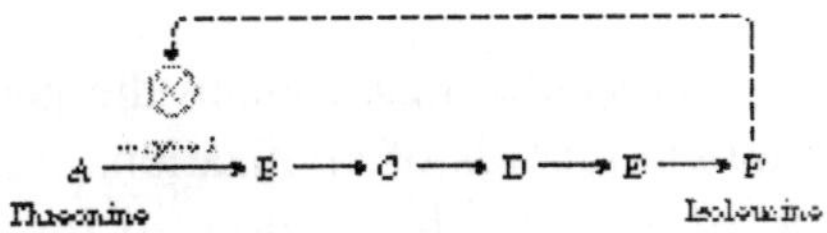

If a cell begins to produce more isoleucine than is needed for protein synthesis, the unused isoleucine accumulates and the increased concentration inhibits the catalytic activity of the first enzyme in the pathway, immediately slowing the production of isoleucine. Such feedback inhibition keeps the production and utilization of each metabolic intermediate in balance.

Although the concept of discrete pathways is an important tool for organizing our understanding of metabolism, it is an oversimplification. There are thousands of metabolic intermediates in a cell, many of which are part of more than one pathway. Metabolism would be better represented as a meshwork of interconnected and interdependent pathways. A change in the concentration of any one metabolite would have an impact on other pathways, starting a ripple effect that would influence the flow of materials through other sectors of the cellular economy.

Cells also regulate the synthesis of their own catalysts, the enzymes, in response to increased or diminished need for a metabolic product. The expression of genes (the translation of information in DNA to active protein in the cell) and synthesis of enzymes are other layers of metabolic control in the cell. All layers must be taken into account when describing the overall control of cellular metabolism. Assembling the complete picture of how the cell regulates itself is a huge job that has only just begun.

Genetic Foundations

Perhaps the most remarkable property of living cells and organisms is their ability to reproduce themselves for countless generations with nearly perfect fidelity. This continuity of inherited traits implies constancy, over millions of years, in the structure of the molecules that contain the genetic information. Very few historical records of civilization, even those etched in copper or carved in stone, have survived for a thousand years. But there is good evidence that the genetic instructions in living organisms have remained nearly unchanged over very much longer periods; many bacteria have nearly the same size, shape, and internal structure and contain the same kinds of precursor molecules and enzymes as bacteria that lived nearly four billion years ago.

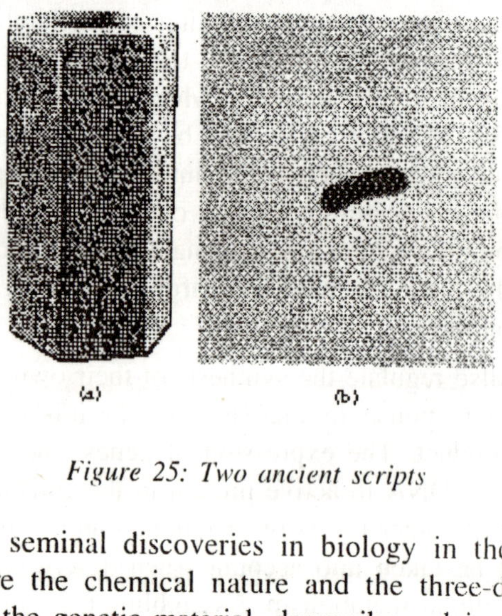

Figure 25: Two ancient scripts

Among the seminal discoveries in biology in the twentieth century were the chemical nature and the three-dimensional structure of the genetic material, deoxyribonucleic acid, DNA. The sequence of the monomeric subunits, the nucleotides , in this linear polymer encodes the instructions for forming all other cellular components and provides a template for the production of identical DNA molecules to be distributed to progeny when a cell divides. The continued existence of a biological species requires its genetic information to be maintained in a stable form, expressed accurately in the form of gene products, and reproduced with a minimum of errors. Effective storage, expression, and reproduction of the genetic message defines individual species, distinguishes them from one another, and assures their continuity over successive generations.

Genetic Continuity

DNA is a long, thin organic polymer, the rare molecule that is constructed on the atomic scale in one dimension (width) and the human scale in another (length: a molecule of DNA can be many centimetres long). A human sperm or egg, carrying the accumulated hereditary information of billions of years of evolution, transmits this inheritance in the form of DNA molecules, in which the linear sequence of covalently linked nucleotide subunits encodes the genetic message.

Usually when we describe the properties of a chemical species, we describe the average behaviour of a very large number of identical molecules. While it is difficult to predict the behaviour of any single molecule in a collection of, say, a picomole (about 6×10^{11} molecules) of a compound, the average behaviour of the molecules is predictable because so many molecules enter into the average. Cellular DNA is a remarkable exception.

The DNA that is the entire genetic material of *E.* coli is a single molecule containing 4.64 million nucleotide pairs. That single molecule must be replicated perfectly in every detail if an *E.* coli cell is to give rise to identical progeny by cell division; there is no room for averaging in this process! The same is true for all cells.

A human sperm brings to the egg that it fertilizes just one molecule of DNA in each of its 23 different chromo-somes, to combine with just one DNA molecule in each corresponding chromosome in the egg. The result of this union is very highly predictable: an embryo with all of its 35,000 genes, constructed of 3 billion nucleotide pairs, intact. An amazing chemical feat!

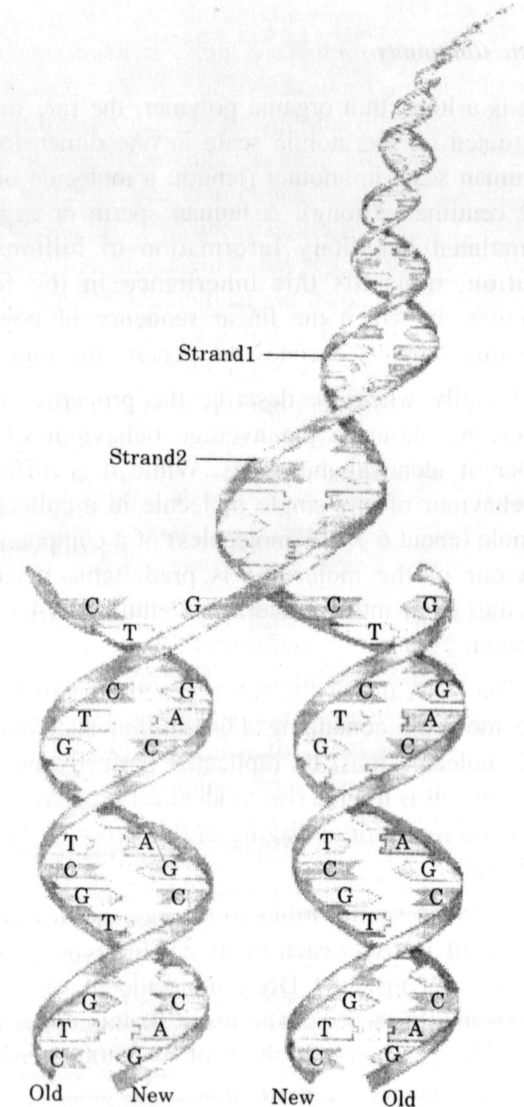

Figure 26: Complementarity between the two strands of DNA.

Replication and Repair with Near-Perfect Fidelity

The capacity of living cells to preserve their genetic material and to duplicate it for the next generation results from the structural complementarity between the two halves of the DNA molecule (Fig. 26). The basic unit of DNA is a linear polymer of four different monomeric subunits, deoxyribonucleotides, arranged in a precise linear sequence. It is this linear sequence that encodes the genetic information. Two of these polymeric strands are twisted about each other to form the DNA double helix, in which each deoxyribonucleotide in one strand pairs specifically with a complementary deoxyribonu-cleotide in the opposite strand. Before a cell divides, the two DNA strands separate and each serves as a template for the synthesis of a new complementary strand, generating two identical double-helical molecules, one for each daughter cell. If one strand is damaged, continuity of information is assured by the information present in the other strand, which acts as a template for repair of the damage.

Linear Sequence in DNA Encodes Proteins

The information in DNA is encoded in its linear (one-dimensional) sequence of deoxyribonucleotide subunits, but the expression of this information results in a three-dimensional cell. This change from one to three dimensions occurs in two phases. A linear sequence of deoxyribonucleotides in DNA codes (through an intermediary, RNA) for the production of a protein with a corresponding linear sequence of amino acids (Fig. 27).

The protein folds into a particular three-dimensional shape, determined by its amino acid sequence and stabilized primarily by noncovalent interactions. Although the final shape of the folded protein is dictated by its amino acid sequence, the folding process is aided by "molecular chaperones," which catalyze the process by discouraging incorrect folding. The precise three-

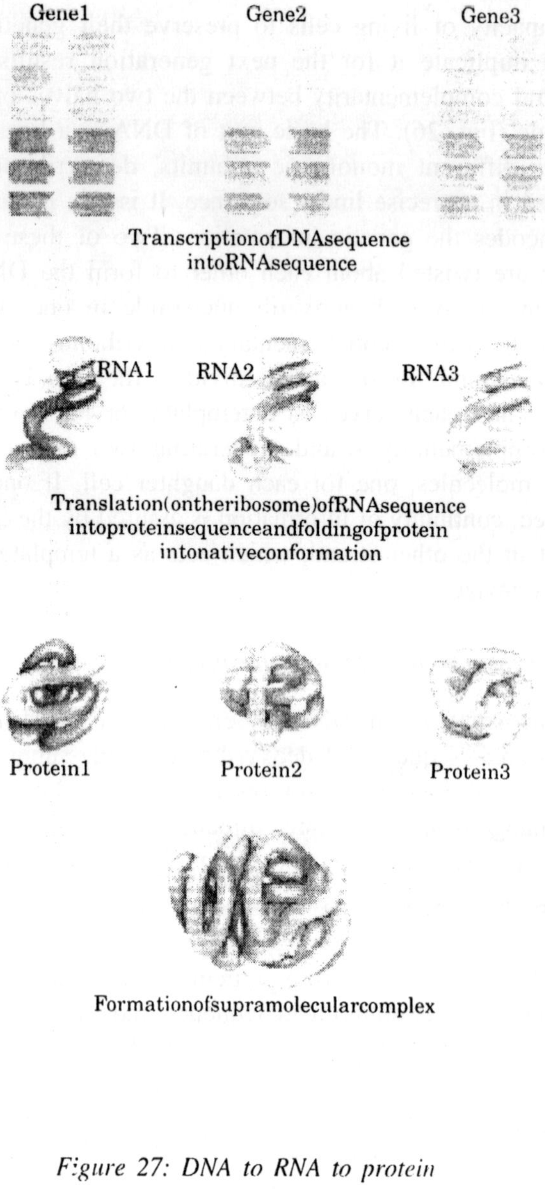

Figure 27: DNA to RNA to protein

dimensiona structure, or native conformation, of the protein is crucial to its function.

Once in its native conformation, a protein may associate noncovalently with other proteins, or with nucleic acids or lipids, to form supramolecular complexes such as chromosomes, ribosomes, and membranes. The individual molecules of these complexes have specific, high-affinity binding sites for each other, and within the cell they spontaneously form functional complexes.

Although protein sequences carry all necessary information for the folding into their native conformation, this correct folding requires the right environment-pH, ionic strength, metal ion concentrations, and so forth. Self-assembly therefore requires both information (provided by the DNA sequence) and environment (the interior of a living cell), and in this sense the DNA sequence alone is not enough to dictate the formation of a cell. As Rudolph Virchow, the nineteenth-century Prussian pathologist and researcher, concluded, "Omnis cellula e cellula": every cell comes from another cell.

Evolutionary Foundations

Great progress in biochemistry and molecular biology during the decades since Dobzhansky made this striking generalization has amply confirmed its validity. The remarkable similarity of metabolic pathways and gene sequences in organisms across the phyla argues strongly that all modern organisms share a common evolutionary progenitor and were derived from it by a series of small changes (mutations), each of which conferred a selective advantage to some organism in some ecological niche.

Changes in the Hereditary Instructions Allow Evolution

Despite the near-perfect fidelity of genetic replication, infrequent, unrepaired mistakes in the DNA replication process lead to changes in the nucleotide sequence of DNA, producing a genetic mutation (Fig. 28) and changing the instructions for some cellular component. Incorrectly repaired damage to one of the DNA strands has the same effect. Mutations in the DNA handed down to offspring-that is, mutations that are carried in the reproductive cells-may be harmful or even lethal to the organism; they may, for example, cause the synthesis of a defective enzyme that is not able to catalyze an essential metabolic reaction. Occasionally, however, a mutation better equips an organism or cell to survive in its environment.

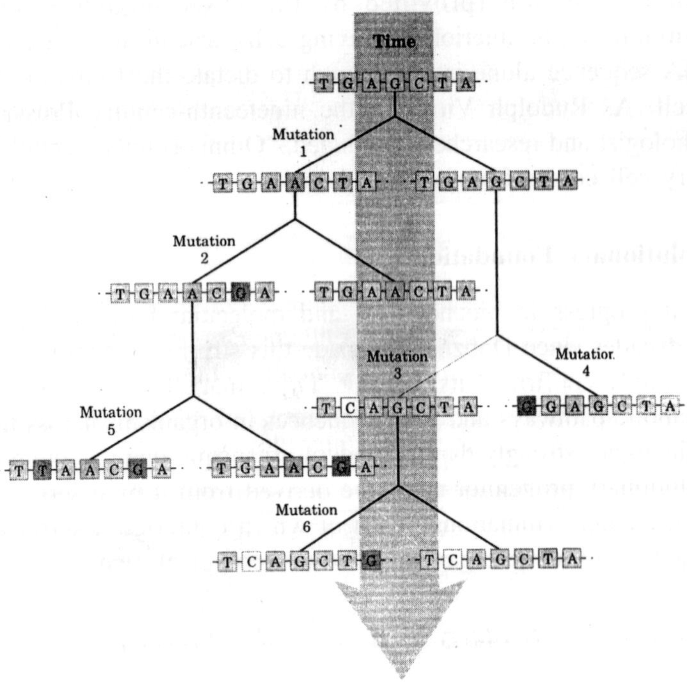

Figure 28: Role of mutation in evolution

The mutant enzyme might have acquired a slightly different specificity, for example, so that it is now able to use some compound that the cell was previously unable to metabolize. If a population of cells were to find itself in an environment where that compound was the only or the most abundant available source of fuel, the mutant cell would have a selective advantage over the other, unmutated (wild-type) cells in the population. The mutant cell and its progeny would survive and prosper in the new environment, whereas wild-type cells would starve and be eliminated. This is what Darwin meant by "survival of the fittest under selective pressure."

Occasionally, a whole gene is duplicated. The second copy is superfluous, and mutations in this gene will not be deleterious; it becomes a means by which the cell may evolve: by producing a new gene with a new function while retaining the original gene and gene function. Seen in this light, the DNA molecules of modern organisms are historical documents, records of the long journey from the earliest cells to modern organisms. The historical accounts in DNA are not complete; in the course of evolution, many mutations must have been erased or written over. But DNA molecules are the best source of biological history that we have.

Several billion years of adaptive selection have refined cellular systems to take maximum advantage of the chemical and physical properties of the molecular raw materials for carrying out the basic energy-transforming and self-replicating activities of a living cell. Chance genetic variations in individuals in a population, combined with natural selection (survival and reproduction of the fittest individuals in a challenging or changing environment), have resulted in the evolution of an enormous variety of organisms, each adapted to life in its particular ecological niche.

Biomolecules First Arose by Chemical Evolution

In 1922, the biochemist Aleksandr I. Oparin proposed a theory for the origin of life early in the history of Earth, postulating that the atmosphere was very different from that of today. Rich in methane, ammonia, and water, and essentially devoid of oxygen, it was a reducing atmosphere, in contrast to the oxidizing environment of our era.

In Oparin's theory, electrical energy from lightning discharges or heat energy from volcanoes caused ammonia, methane, water vapor, and other components of the primitive atmosphere to react, forming simple organic compounds. These compounds then dissolved in the ancient seas, which over many millennia became enriched with a large variety of simple organic substances. In this warm solution (the "primordial soup"), some organic molecules had a greater tendency than others to associate into larger complexes. Over millions of years, these in turn assembled spontaneously to form membranes and catalysts (enzymes), which came together to become precursors of the earliest cells. Oparin's views remained speculative for many years and appeared untestable-until a surprising experiment was conducted using simple equipment on a desktop.

Chemical Evolution Can Be Simulated in the Laboratory

The classic experiment on the abiotic (nonbiological) origin of organic biomolecules was carried out in 1953 by Stanley Miller in the laboratory of Harold Urey. Miller subjected gaseous mixtures of NH_3, CH_4, H_2O, and H_2 to electrical sparks produced across a pair of electrodes (to simulate lightning) for periods of a week or more, then analysed the contents of the closed reaction vessel (Fig. 29). The gas phase of the resulting mixture contained CO and CO_2, as well as the starting materials. The water phase contained a variety of organic compounds, including some amino acids, hydroxy acids, aldehydes, and hydrogen cyanide (HCN). This experiment established the possibility of abiotic production of biomolecules in relatively short times under relatively mild conditions.

More refined laboratory experiments have provided good evidence that many of the chemical components of living cells, including polypeptides and RNA-like molecules, can form under these conditions. Polymers of RNA can act as catalysts in biologically significant reactions and RNA probably played a crucial role in prebiotic evolution, both as catalyst and as information repository.

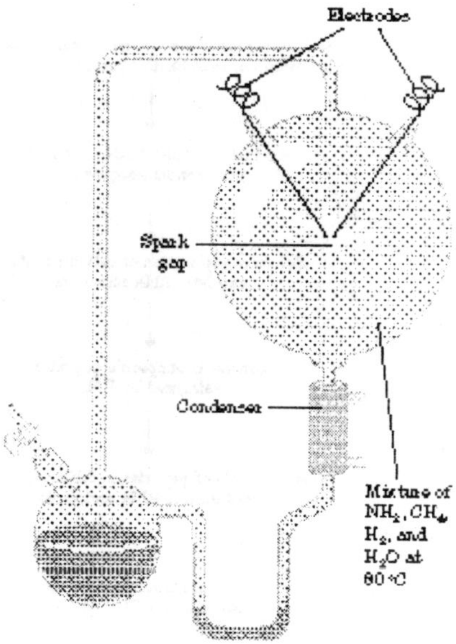

Figure 29: Abiotic production of biomolecules.

RNA or Related Precursors

In modern organisms, nucleic acids encode the genetic information that specifies the structure of enzymes, and enzymes catalyze the replication and repair of nucleic acids. The mutual dependence of these two classes of biomolecules brings up the perplexing question: which came first, DNA or protein?

The discovery that RNA molecules can act as catalysts in their own formation suggests that RNA or a similar molecule may have been the first gene and the first catalyst. According to this scenario (Fig. 30), one of the earliest stages of biological evolution was the chance formation, in the primordial soup, of an RNA molecule that could catalyze the formation of other RNA molecules of the same sequence—a self-replicating, self-perpetuating RNA.

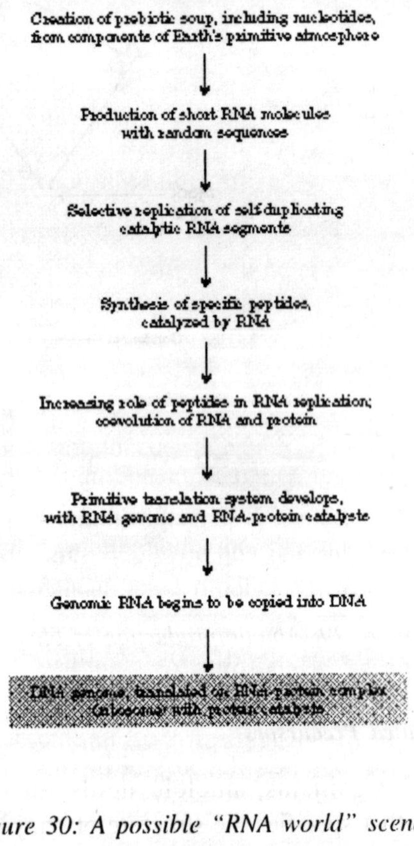

Figure 30: A possible "RNA world" scenario

The concentration of a self-replicating RNA molecule would

increase exponentially, as one molecule formed two, two formed four, and so on. The fidelity of self-replication was presumably less than perfect, so the process would generate variants of the RNA, some of which might be even better able to self-replicate. In the competition for nucleotides, the most efficient of the self-replicating sequences would win, and less efficient replicators would fade from the population.

The division of function between DNA (genetic information storage) and protein (catalysis) was, according to the "RNA world" hypothesis, a later development. New variants of self-replicating RNA molecules developed, with the additional ability to catalyze the condensation of amino acids into peptides. Occasionally, the peptide(s) thus formed would reinforce the self-replicating ability of the RNA, and the pair-RNA molecule and helping peptide-could undergo further modifications in sequence, generating even more efficient self-replicating systems. The recent, remarkable discovery that, in the protein-synthesizing machinery of modern cells (ribosomes), RNA molecules, not proteins, catalyze the formation of peptide bonds is certainly consistent with the RNA world hypothesis. Some time after the evolution of this primitive protein-synthesizing system, there was a further development:

DNA molecules with sequences complementary to the self-replicating RNA molecules took over the function of conserving the "genetic" information, and RNA molecules evolved to play roles in protein synthesis. Proteins proved to be versatile catalysts and, over time, took over that function. Lipidlike compounds in the primordial soup formed relatively impermeable layers around self-replicating collections of molecules. The concentration of proteins and nucleic acids within these lipid enclosures favoured the molecular interactions required in self-replication.

2

Biochemistry of Water

Water is the most abundant substance in living systems, making up 70% or more of the weight of most organisms. The first living organisms doubtless arose in an aqueous environment, and the course of evolution has been shaped by the properties of the aqueous medium in which life began. The attractive forces between water molecules and the slight tendency of water to ionize are of crucial importance to the structure and function of biomolecules. The water molecule and its ionization products, H^+ and OH^-, profoundly influence the structure, self-assembly, and properties of all cellular components, including proteins, nucleic acids, and lipids. The noncovalent interactions responsible for the strength and specificity of "recognition" among biomolecules are decisively influenced by the solvent properties of water, including its ability to form hydrogen bonds with itself and with solutes.

Weak Interactions in Aqueous Systems

Hydrogen bonds between water molecules provide the cohesive forces that make water a liquid at room temperature and that favor the extreme ordering of molecules that is typical of crystalline water (ice). Polar biomolecules dissolve readily in water because they can replace water-water interactions with more energetically favorable water-solute interactions. In contrast, nonpolar biomolecules interfere with water-water interactions but are unable to form water-solute interactions-consequently, nonpolar molecules are poorly soluble in water. In aqueous solutions, nonpolar molecules tend to cluster together.

Hydrogen bonds and ionic, hydrophobic (Greek, "water-fearing"), and van der Waals interactions are individually weak, but collectively they have a very significant influence on the three-dimensional structures of proteins, nucleic acids, polysaccharides, and membrane lipids.

Water has a higher melting point, boiling point, and heat of vaporization than most other common solvents . These unusual properties are a consequence of attractions between adjacent water molecules that give liquid water great internal cohesion. A look at the electron structure of the H_2O molecule reveals the cause of these intermolecular attractions.

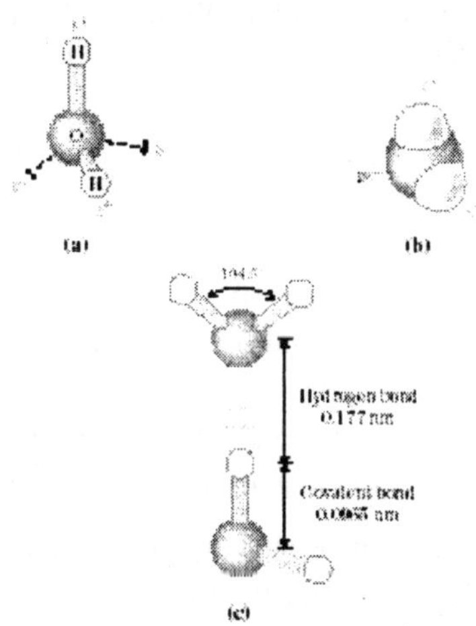

Figure 1: Structure of the water molecule

In Figure 1 the dipolar nature of the H_2O molecule is shown by (a) ball-and-stick and (b) space-filling models. The dashed lines in (a) represent the nonbonding orbitals. There is a nearly tetrahedral arrangement of the outer-shell electron pairs around the oxygen atom; the two hydrogen atoms have localized partial positive charges (d⁺) and the oxygen atom has a partial negative charge (2d⁻). (c) Two H_2O molecules joined by a hydrogen bond between the oxygen atom of the upper molecule and a hydrogen atom of the lower one. Hydrogen bonds are longer and weaker than covalent O—H bonds.

Each hydrogen atom of a water molecule shares an electron pair with the central oxygen atom. The geometry of the molecule is dictated by the shapes of the outer electron orbitals of the oxygen atom, which are similar to the sp_3 bon ding orbitals of carbon. These orbitals describe a rough tetrahedron, with a hydrogen atom at each of two corners and unshared electron pairs at the other two corners. The H—O—H bond angle is 104.5°, slightly less than the 109.5° of a perfect tetrahedron because of crowding by the nonbonding orbitals of the oxygen atom.

The oxygen nucleus attracts electrons more strongly than does the hydrogen nucleus (a proton); that is, oxygen is more electronegative. The sharing of electrons between H and O is therefore unequal; the electrons are more often in the vicinity of the oxygen atom than of the hydrogen. The result of this unequal electron sharing is two electric dipoles in the water molecule, one along each of the H—O bonds; each hydrogen bears a partial positive charge (δ^+) and the oxygen atom bears a partial negative charge equal to the sum of the two partial positives ($2\delta^-$). As a result, there is an electrostatic attraction between the oxygen atom of one water molecule and the hydrogen of another, called a hydrogen bond.

Hydrogen bonds are relatively weak. Those in liquidwater have a bond dissociation energy (the energy required to break a bond) of about 23 kJ/mol, compared with 470 kJ/mol for the covalent O—H bond in water or 348 kJ/mol for a covalent C—C bond. The hydrogen bond is about 10% covalent, due to overlaps in the bonding orbitals, and about 90% electrostatic.

At room temperature, the thermal energy of an aqueous solution is of the same order of magnitude as that required to break hydrogen bonds. When water is heated, the increase in temperature reflects the faster motion of individual water molecules. At any given time, most of the molecules in liquid water are engaged in hydrogen bonding, but the lifetime of each hydrogen bond is just 1 to 20 picoseconds (1 ps = 10^{-12} s); upon breakage of one hydrogen bond, another hydrogen bond forms, with the same partner or a new one, within 0.1 ps.

The apt phrase "flickering clusters" has been applied to the short-lived groups of water molecules interlinked by hydrogen bonds in liquid water. The sum of all the hydrogen bonds between H_2O molecules confers great internal cohesion on liquid water. Extended networks of hydrogen-bonded water molecules also form bridges between solutes that allow the larger molecules to interact with each other over distances of several nanometers without physically touching.

The nearly tetrahedral arrangement of the orbitals about the oxygen atom allows each water molecule to form hydrogen bonds with as many as four neighboring water molecules. In liquid water at room temperature and atmospheric pressure, however, water molecules are disorganized and in continuous motion, so that each molecule forms hydrogen bonds with an average of only 3.4 other molecules. In ice, on the other hand, each water molecule is fixed in space and forms hydrogen bonds with a full complement of four other water molecules to yield a regular lattice structure. Breaking a sufficient proportion of hydrogen bonds to destabilize the crystal lattice of ice requires much thermal energy, which accounts for the relatively high melting point of water. When ice melts or water evaporates, heat is taken up by the system:

$$H_2O(solid) \rightarrow H_2O(liquid) \quad \Delta H = \pm 5.9 \text{ kJ/mol}$$

$$H_2O(liquid) \rightarrow H_2O(gas) \quad \Delta H = \pm 44.0 \text{ kJ/mol}$$

During melting or evaporation, the entropy of the aqueous system increases as more highly ordered arrays of water molecules relax into the less orderly hydrogenbonded arrays in liquid water or the wholly disordered gaseous state.

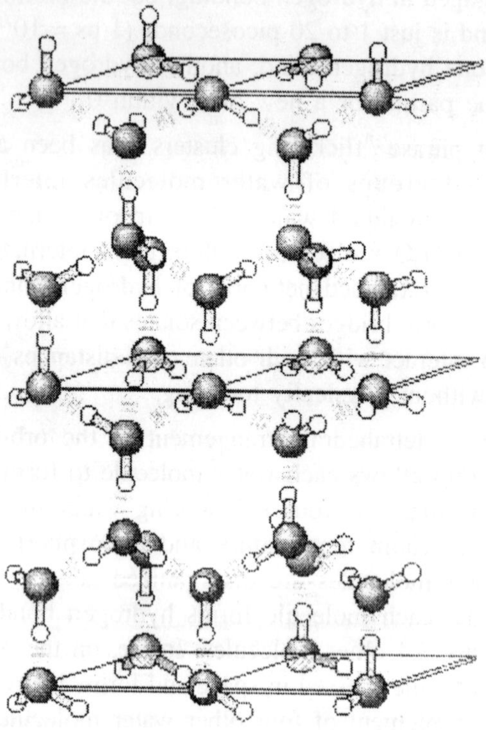

Figure 2: Hydrogen bonding in ice

At room temperature, both the melting of ice and the evaporation of water occur spontaneously; the tendency of the water molecules to associate through hydrogen bonds is outweighed by the energetic push toward randomness. Recall that the free-energy change (ΔG) must have a negative value for a process to occur spontaneously: $\Delta G = \Delta H - T\,\Delta S$, where ΔG represents the driving force, ΔH the enthalpy change from making and breaking bonds, and ΔS the change in randomness. Because ΔH

is positive for melting and evaporation, it is clearly the increase in entropy (ΔS) that makes ΔG negative and drives these transformations.

Water Forms Hydrogen Bonds with Polar Solutes

Hydrogen bonds are not unique to water. They readily form between an electronegative atom (the hydrogen acceptor, usually oxygen or nitrogen with a lone pair of electrons) and a hydrogen atom covalently bonded to another electronegative atom (the hydrogen donor) in the same or another molecule.

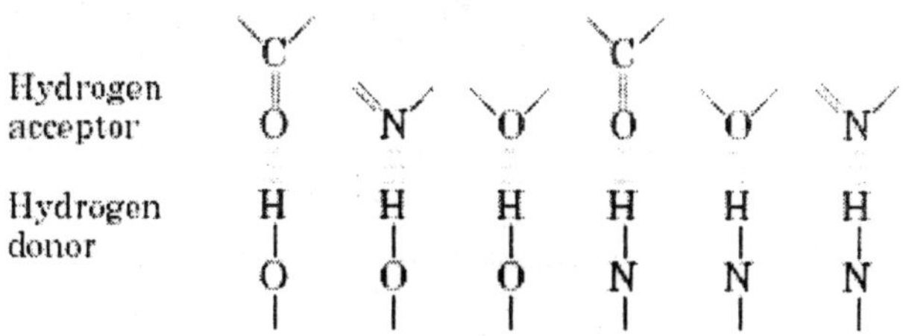

Figure 3: Common hydrogen bonds in biological systems.

Hydrogen atoms covalently bonded to carbon atoms do not participate in hydrogen bonding, because carbon is only slightly more electronegative than hydrogen and thus the C—H bond is only very weakly polar. The distinction explains why butanol ($CH_3(CH_2)_2CH_2OH$) has a relatively high boiling point of 117°C, whereas butane ($CH_3(CH_2)_2CH_3$) has a boiling point of only –0.5°C.

Butanol has a polar hydroxyl group and thus can form intermolecular hydrogen bonds. Uncharged but polar biomolecules such as sugars dissolve readily in water because of the stabilizing effect of hydrogen bonds between the hydroxyl groups or carbonyl oxygen of the sugar and the polar water molecules. Alcohols, aldehydes, ketones, and compounds containing N—H bonds all form hydrogen bonds with water molecules and tend to be soluble in water.

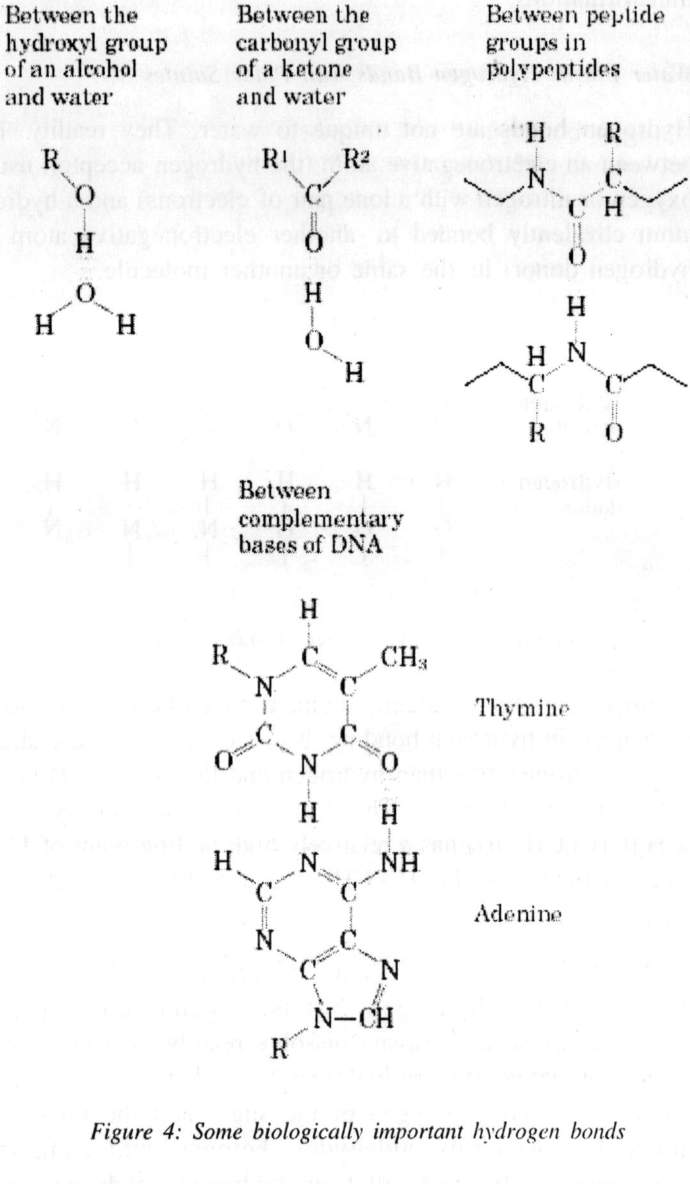

Figure 4: Some biologically important hydrogen bonds

Hydrogen bonds are strongest when the bonded molecules are oriented to maximize electrostatic interaction, which occurs when the hydrogen atom and the two atoms that share it are in a straight line-that is, when the acceptor atom is in line with the covalent bond between the donor atom and H. Hydrogen bonds are thus highly directional and capable of holding holding two hydrogen-bonded molecules or groups in a specific geometric arrangement. This property of hydrogen bonds confers very precise threedimensional structures on protein and nucleic acid molecules, which have many intramolecular hydrogen bonds.

Water Interacts Electrostatically with Charged Solutes

Water is a polar solvent. It readily dissolves most biomolecules, which are generally charged or polar compounds ; compounds that dissolve easily in water are hydrophilic (Greek, "water-loving"). In contrast, nonpolar solvents such as chloroform and benzene are poor solvents for polar biomolecules but easily dissolve those that are hydrophobic-nonpolar molecules such as lipids and waxes.

Water dissolves salts such as NaCl by hydrating and stabilizing the Na^+ and Cl^- ions, weakening the electrostatic interactions between them and thus counteracting their tendency to associate in a crystalline lattice. The same factors apply to charged biomolecules, compounds with functional groups such as ionized carboxylic acids (—COO^-), protonated amines(—NH_3^+), and phosphate esters or anhydrides. Water readily dissolves such compounds by replacing solutesolute hydrogen bonds with solute-water hydrogen bonds, thus screening the electrostatic interactions between solute molecules.

Water is especially effective in screening the electrostatic interactions between dissolved ions because it has a high dielectric constant, a physical property reflecting the number of dipoles in a solvent. The strength, or force (F), of ionic interactions in a solution depends upon the magnitude of the charges (Q), the distance between the charged groups(γ), and the dielectric constant (ε) of the solvent in which the interactions occur:

$$F = \frac{Q_1 Q_2}{\varepsilon r^2}$$

Table 1: Examples of Polar, Nonpolar, and Ammphipathic Biomolecules

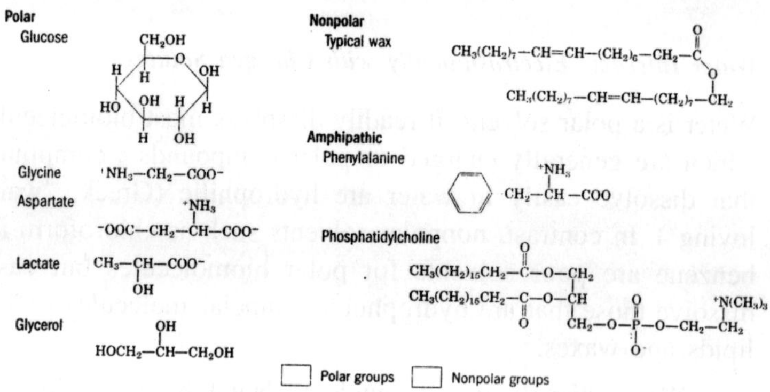

For water at 25°C, ε (which is dimensionless) is 78.5, and for the very nonpolar solvent benzene, ε is 4.6. Thus, ionic interactions are much stronger in less polar environments. The dependence on r^2 is such that ionic attractions or repulsions operate only over short distances in the range of 10 to 40 nm (depending on the electrolyte concentration) when the solvent is water.

Entropy Increases as Crystalline Substances Dissolve

As a salt such as NaCl dissolves, the Na^+ and Cl^- ions leaving the crystal lattice acquire far greater freedom of motion. The resulting increase in entropy (randomness) of the system is largely responsible for the ease of dissolving salts such as NaCl in water. In thermodynamic terms, formation of the solution occurs with a favorable free-energy change: $\Delta G = \Delta H - T\Delta S$, where ΔH has a small positive value and $T\Delta S$ a large positive value; thus ΔG is negative.

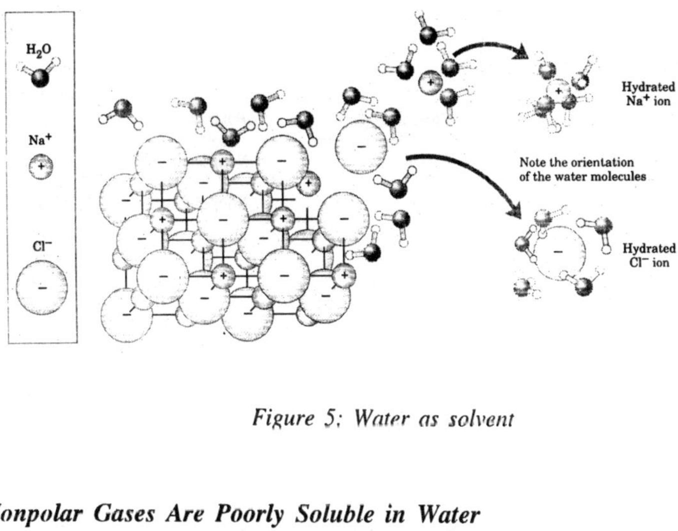

Figure 5: Water as solvent

Nonpolar Gases Are Poorly Soluble in Water

The molecules of the biologically important gases CO_2, O_2, and N_2 are nonpolar. In O_2 and N_2, electrons are shared equally by both atoms. In CO_2, each C=O bond is polar, but the two dipoles are oppositely directed and cancel each other. The movement of molecules from the disordered gas phase into aqueous solution constrains their motion and the motion of water molecules and therefore represents a decrease in entropy. The nonpolar nature of these gases and the decrease in entropy when they enter solution combine to make them very poorly soluble in water . Some organisms have water-soluble carrier proteins (hemoglobin and myoglobin, for example) that facilitate the transport of O_2. Carbon dioxide forms carbonic acid (H_2CO_3) in aqueous solution and is transported as the HCO_3^-(bicarbonate) ion, either free-bicarbonate is very soluble in water (~100 g/L at 25°C)—or bound to hemoglobin. Two other gases, NH_3 and H_2S, also have biological roles in some organisms; these gases are polar and dissolve readily in water.

Gas	Structure*	Polarity	Solubility in water (g/L)†
Nitrogen	N≡N	Nonpolar	0.018 (40 °C)
Oxygen	O=O	Nonpolar	0.035 (50 °C)
Carbon dioxide	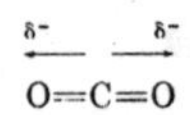	Nonpolar	0.97 (45 °C)
Ammonia	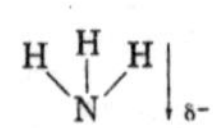	Polar	900 (10 °C)
Hydrogen sulfide	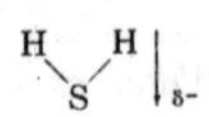	Polar	1,860 (40 °C)

*The arrows represent electric dipoles; there is a partial negative charge (δ^-) at the head of the arrow, a partial positive charge (δ^+; not shown here) at the tail.

†Note that polar molecules dissolve far better even at low temperatures than do nonpolar molecules at relatively high temperatures.

Table 2: Solubility of some gases in water

Nonpolar Compounds Force Energetically Unfavorable Changes in the Structure of Water

When water is mixed with benzene or hexane, two phases form; neither liquid is soluble in the other. Nonpolar compounds such as benzene and hexane are hydrophobic-they are unable to undergo energetically favorable interactions with water molecules, and they interfere with the hydrogen bonding among water molecules. All molecules or ions in aqueous solution interfere with the hydrogen bonding of some water molecules in their immediate vicinity, but polar or charged solutes (such as NaCl) compensate for lost water-water hydrogen bonds by forming new solute-water interactions.

The net change in enthalpy (ΔH) for dissolving these solutes is generally small. Hydrophobic solutes, however, offer no such compensation, and their addition to water may therefore result in a small gain of enthalpy; the breaking of hydrogen bonds between water molecules takes up energy from the system. Furthermore, dissolving hydrophobic compounds in water produces a measurable decrease in entropy.

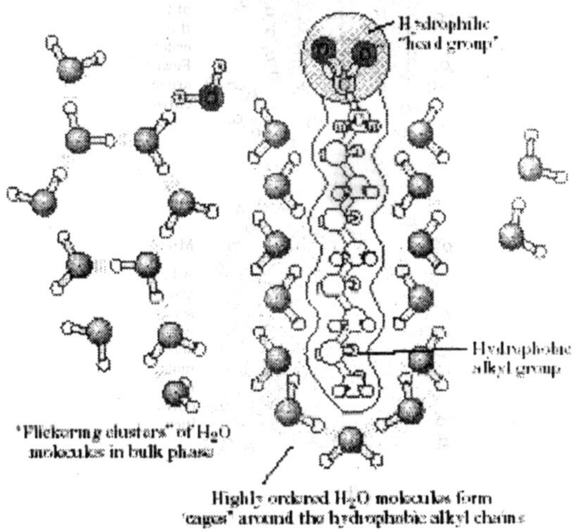

(a)

Figure 6: Amphipathic compounds in aqueous solution. (a)Longchain fatty acids have very hydrophobic alkyl chains, each of which is surrounded by a layer of highly ordered water molecules.

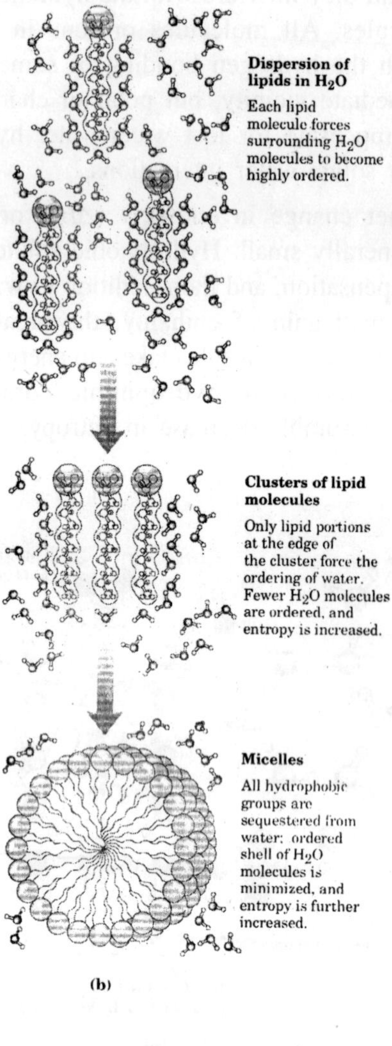

Figure 6(b): By clustering together in micelles, the fatty acid molecules expose the smallest possible hydrophobic surface area to the water, and fewer water molecules are required in the shell of ordered water.

Water molecules in the immediate vicinity of a nonpolar solute are constrained in their possible orientations as they form a highly ordered cagelike shell around each solute molecule. These water molecules are not as highly oriented as those in clathrates, crystalline compounds of nonpolar solutes and water, but the effect is the same in both cases: the ordering of water molecules reduces entropy. The number of ordered water molecules, and therefore the magnitude of the entropy decrease, is proportional to the surface area of the hydrophobic solute enclosed within the cage of water molecules. The freeenergy change for dissolving a nonpolar solute in water is thus unfavorable: $DG=DH-TDS$, where DH has a positive value, DS has a negative value, and DG is positive.

Amphipathic compounds contain regions that are polar (or charged) and regions that are nonpolar (Table 1). When an amphipathic compound is mixed with water, the polar, hydrophilic region interacts favorably with the solvent and tends to dissolve, but the nonpolar, hydrophobic region tends to avoid contact with the water (Fig. 6a). The nonpolar regions of the molecules cluster together to present the smallest hydrophobic area to the aqueous solvent, and the polar regions are arranged to maximize their interaction with the solvent (Fig. 6b).

These stable structures of amphipathic compounds in water, called micelles, may contain hundreds or thousands of molecules. The forces that hold the nonpolar regions of the molecules together are called hydrophobic interactions. The strength of hydrophobic interactions is not due to any intrinsic attraction between nonpolar moieties. Rather, it results from the system's achieving greatest thermodynamic stability by minimizing the number of ordered water molecules required to surround hydrophobic portions of the solute molecules.

Many biomolecules are amphipathic; proteins, pigments, certain vitamins, and the sterols and phospholipids of membranes all have polar and nonpolar surface regions. Structures composed of these molecules are stabilized by hydrophobic interactions among the nonpolar regions.

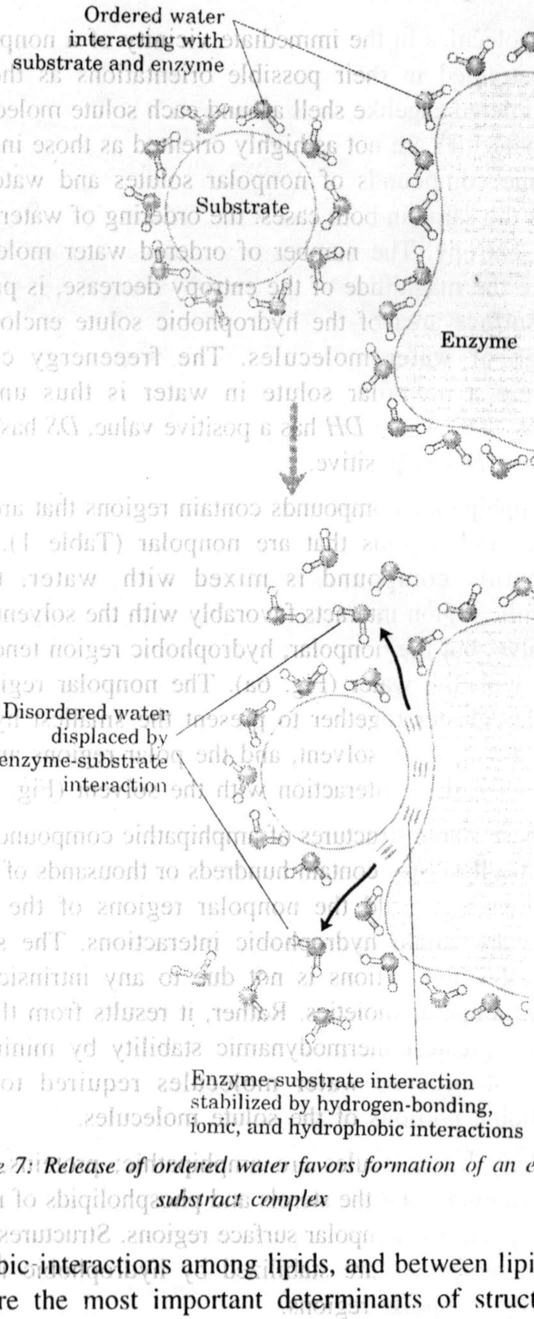

Figure 7: Release of ordered water favors formation of an enzyme-substract complex

Hydrophobic interactions among lipids, and between lipids and proteins, are the most important determinants of structure in biological membranes. Hydrophobic interactions between nonpolar amino acids also stabilize the three-dimensional structures of proteins.

Hydrogen bonding between water and polar solutes also causes some ordering of water molecules, but the effect is less significant than with nonpolar solutes. Part of the driving force for binding of a polar substrate (reactant) to the complementary polar surface of an enzyme is the entropy increase as the enzyme displaces ordered water from the substrate (Fig. 7).

van der Waals Interactions Are Weak Interatomic Attractions

When two uncharged atoms are brought very close together, their surrounding electron clouds influence each other. Random variations in the positions of the electrons around one nucleus may create a transient electric dipole, which induces a transient, opposite electric dipole in the nearby atom. The two dipoles weakly attract each other, bringing the two nuclei closer. These weak attractions are called van der Waals interactions.

Table 3: van der Waals radii and Covalent radii of some elements

Element	van der Waals radius (nm)	Covalent radius for single bond (nm)
H	0.11	0.030
O	0.15	0.066
N	0.15	0.070
C	0.17	0.077
S	0.18	0.104
P	0.19	0.110
I	0.21	0.133

As the two nuclei draw closer together, their electron clouds begin to repel each other. At the point where the van der Waals attraction exactly balances this repulsive force, the nuclei are said to be in van der Waals contact. Each atom has a characteristic van der Waals radius, a measure of how close that atom will allow another to approach (Table 3).

The noncovalent interactions we have described (hydrogen bonds and ionic, hydrophobic, and van der Waals interactions) (Table 4) are much weaker than covalent bonds. An input of about 350 kJ of energy is required to break a mole of (6×10^{23}) C—C single bonds, and about 410 kJ to break a mole of C—H bonds, but as little as 4 kJ is sufficient to disrupt a mole of typical van der Waals interactions.

Table 4: Four types of noncovalent interactions among biomolecules in aqueous solvent

Hydrogen bonds	
Between neutral groups	>C=O···H—O—
Between peptide bonds	>C=O···H—N<
Ionic interactions	
Attraction	$—^{+}NH_3 \rightarrow\leftarrow {}^{-}O—C(=O)—$
Repulsion	$—^{+}NH_3 \leftarrow\rightarrow H_3N^{+}—$
Hydrophobic interactions	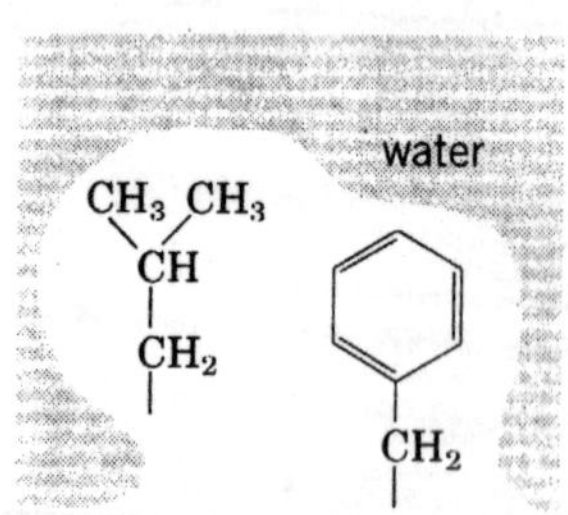
van der Waals interactions	Any two atoms in close proximity

Hydrophobic interactions are also much weaker than covalent bonds, although they are substantially strengthened by a highly polar solvent (a concentrated salt solution, for example). Ionic interactions and hydrogen bonds are variable in strength, depending on the polarity of the solvent and the alignment of the hydrogen-bonded atoms, but they are always significantly weaker than covalent bonds.

In aqueous solvent at 25°C, the available thermal energy can be of the same order of magnitude as the strength of these weak interactions, and the interaction between solute and solvent (water) molecules is nearly as favorable as solute-solute interactions. Consequently, hydrogen bonds and ionic, hydrophobic, and van der Waals interactions are continually formed and broken.

Although these four types of interactions are individually weak relative to covalent bonds, the cumulative effect of many such interactions can be very significant.For example, the noncovalent binding of an enzyme toits substrate may involve several hydrogen bonds and one or more ionic interactions, as well as hydrophobic and van der Waals interactions. The formation of each of these weak bonds contributes to a net decrease in the free energy of the system. We can calculate the stability of a noncovalent interaction, such as that of a small molecule hydrogen-bonded to its macromolecular partner, from the binding energy. Stability, as measured by the equilibrium constant of the binding reaction, varies exponentially with binding energy.

The dissociation of two biomolecules (such as an enzyme and its bound substrate) associated noncovalently through multiple weak interactions requires all these interactions to be disrupted at the same time. Because the interactions fluctuate randomly, such simultaneous disruptions are very unlikely. The molecular stability bestowed by 5 or 20 weak interactions is therefore much greater than would be expected intuitively from a simple summation of small binding energies.

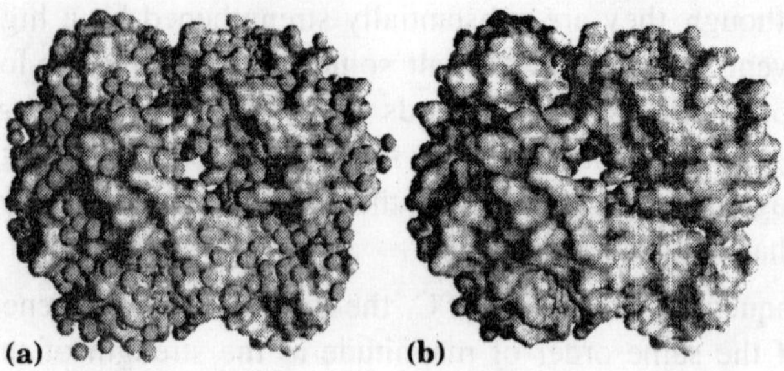

Figure 8: Water binding in hemoglobin. (a) The crystal structure of hemoglobin with water molecules and (b) without the water molecules.

Macromolecules such as proteins, DNA, and RNA contain so many sites of potential hydrogen bonding or ionic, van der Waals, or hydrophobic interactions that the cumulative effect of the many small binding forces can be enormous. For macromolecules, the most stable (that is, the native) structure is usually that in which weak-bonding possibilities are maximized. The folding of a single polypeptide or polynucleotide chain into its three-dimensional shape is determined by this principle.

The binding of an antigen to a specific antibody depends on the cumulative effects of many weak interactions. The energy released when an enzyme binds noncovalently to its substrate is the main source of the enzyme's catalytic power. The binding of a hormone or a neurotransmitter to its cellular receptor protein is the result of weak interactions. One consequence of the large size of enzymes and receptors is that their extensive surfaces provide many opportunities for weak interactions. At the molecular level, the complementarity between interacting biomolecules reflects the complementarity and weak interactions between polar, charged, and hydrophobic groups on the surfaces of the molecules.

When the structure of a protein such as hemoglobin (Fig.8) is determined by x-ray crystallography, water molecules are often found to be bound so tightly as to be part of the crystal structure; the same is true for water in crystals of RNA or DNA. These bound water molecules, which can also be detected in aqueous solutions by nuclear magnetic resonance, have distinctly different properties from those of the "bulk" water of the solvent. They are, for example, not osmotically active. For many proteins, tightly bound water molecules are essential to their function. In a reaction central to the process of photosynthesis, for example, light drives protons across a biological membrane as electrons flow through a series of electron-carrying proteins

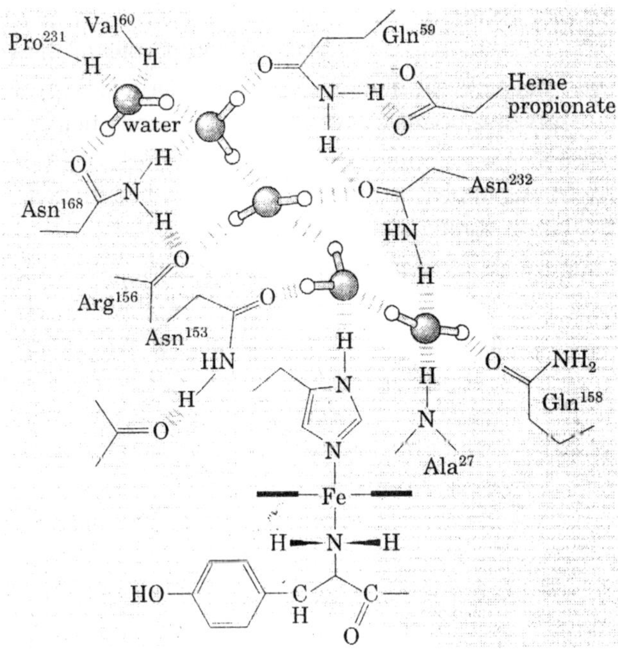

Figure 9: Water chain in cytochrome f

One of these proteins, cytochrome f, has a chain of five bound water molecules (Fig. 9) that may provide a path for protons to move through the membrane by a process known as "proton hopping". Another such light-driven proton pump.

bacteriorhodopsin, almost certainly uses a chain of precisely oriented bound water molecules in the transmembrane movement of protons.

Solutes Affect the Colligative Properties of Aqueous Solutions

Solutes of all kinds alter certain physical properties of the solvent, water: its vapor pressure, boiling point, melting point (freezing point), and osmotic pressure. These are called colligative ("tied together") properties, because the effect of solutes on all four properties has the same basis: the concentration of water is lower in solutions than in pure water. The effect of solute concentration on the colligative properties of water is independent of the chemical properties of the solute; it depends only on the number of solute particles (molecules, ions) in a given amount of water. A compound such as NaCl, which dissociates in solution, has twice the effect on osmotic pressure, for example, as does an equal number of moles of a nondissociating solute such as glucose.

Solutes alter the colligative properties of aqueous solutions by lowering the effective concentration of water. For example, when a significant fraction of the molecules at the surface of an aqueous solution are not water but solute, the tendency of water molecules to escape into the vapor phase-that is, the vapor pressure is lowered (Fig. 10). Similarly, the tendency of water molecules to move from the aqueous phase to the surface of a forming ice crystal is reduced when some of the molecules that collide with the crystal are solute, not water. In that case, the solution will freeze more slowly than pure water and at a lower temperature.

For a 1.00 molal aqueous solution (1.00 mol of solute per 1,000 g of water) of an ideal, nonvolatile, and nondissociating solute at 101 kPa (1 atm) of pressure, the freezing point is 1.86°C lower and the boiling point is 0.543° C higher than for pure water. For a 0.100 molal solution of the same solute, the changes are one-tenth as large.

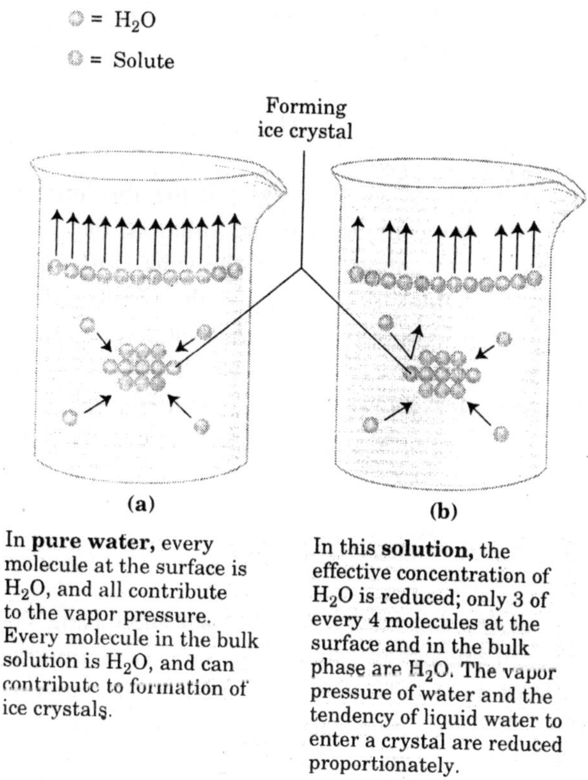

Figure 10: Solutes alter the colligative properties of aqueous solutions. (a) At 101 kPa (1 atm) pressure, pure water boils at 100 °C and freezes at 0 °C. (b) The presence of solute molecules reduces the probability of a water molecule leaving the solution and entering the gas phase, thereby reducing the vapor pressure of the solution and increasing the boiling point.

Water molecules tend to move from a region of higher water concentration to one of lower water concentration. When two different aqueous solutions are separated by a semipermeable membrane (one that allows the passage of water but not solute molecules), water molecules diffusing from the region of higher water concentration to that of lower water concentration produce osmotic pressure (Fig.11). This pressure, , measured as the force necessary to resist water movement (Fig. 11c), is approximated by the van't Hoff equation:

$$\Pi = icRT$$

in which R is the gas constant and T is the absolute temperature. The term ic is the osmolarity of the solution, the product of the solute's molar concentration c and the van't Hoff factor i, which is a measure of the extent to which the solute dissociates into two or more ionic species. In dilute NaCl solutions, the solute completely dissociates into Na^+ and Cl^-, doubling the number of solute particles, and thus $i = 2$. For nonionizing solutes, i is always 1. For solutions of several (n) solutes, Π is the sum of the contributions of each species:

$$\Pi = RT(i_1c_1 + i_2c_2 + \dots + i_nc_n)$$

Osmosis, water movement across a semipermeable membrane driven by differences in osmotic pressure, is an important factor in the life of most cells. Plasma membranes are more permeable to water than to most other small molecules, ions, and macromolecules. This permeability is due partly to simple diffusion of water through the lipid bilayer and partly to protein channels in the membrane that selectively permit the passage of water. Solutions of equal osmolarity are said to be isotonic. Surrounded by an isotonic solution, a cell neither gains nor loses water (Fig.12). In a hypertonic solution, one with higher

osmolarity than the cytosol, the cell shrinks as water flows out. In a hypotonic solution, with lower osmolarity than the cytosol, the cell swells as water enters. In their natural environments, cells generally contain higher concentrations of biomolecules and ions than their surroundings, so osmotic pressure tends to drive water into cells. If not somehow counterbalanced, this inward movement of water would distend the plasma membrane and eventually cause bursting of the cell (osmotic lysis).

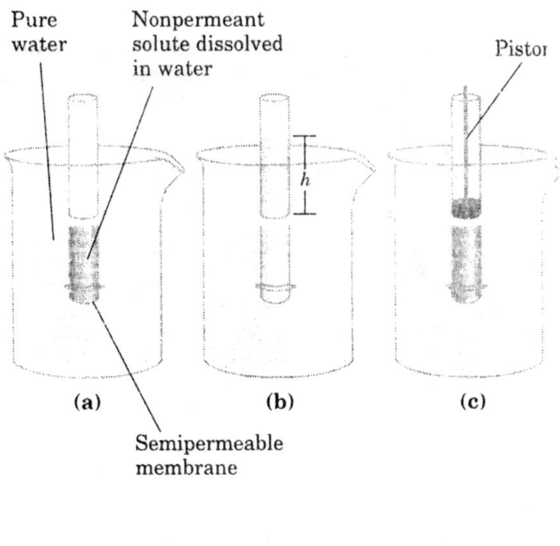

Figure 11: Osmosis and the measurement of osmotic pressure. (a) The initial state. (b) The final state. (c) Osmotic pressure (Π) is measured as the force that must be applied to return the solution in the tube to the level of that in the beaker.

Several mechanisms have evolved to prevent this catastrophe. In bacteria and plants, the plasma membrane is surrounded by a nonexpandable cell wall of sufficient rigidity and strength to resist osmotic pressure and prevent osmotic lysis. Certain freshwater protists that live in a highly hypotonic medium have an organelle (contractile vacuole) that pumps water out of the cell. In multicellular animals, blood plasma and interstitial fluid (the extracellular fluid of tissues) are maintained at an osmolarity close to that of the cytosol. The high concentration of albumin and other proteins in blood plasma contributes to its osmolarity. Cells also actively pump out ions such as Na^+ into the interstitial fluid to stay in osmotic balance with their surroundings.

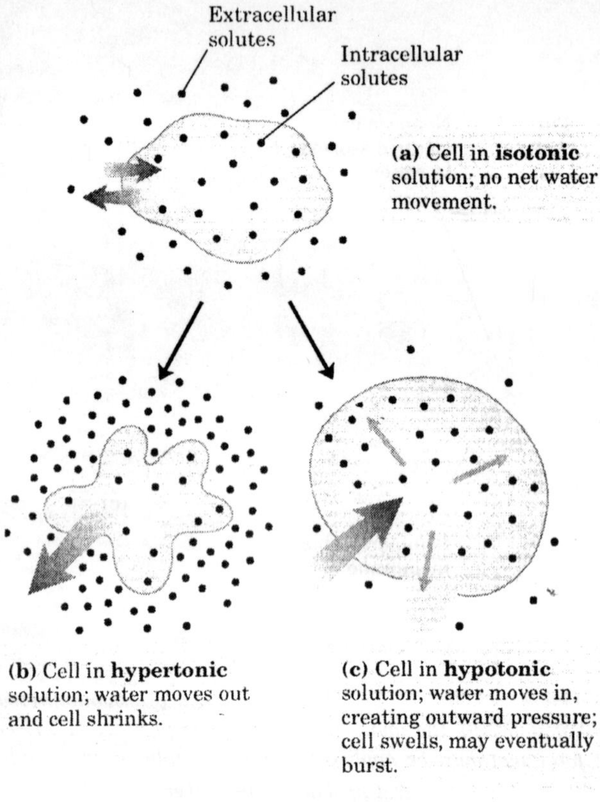

Figure 12:Effect of extracellular osmolarity on water movement across a plasma membrane.

Because the effect of solutes on osmolarity depends on the number of dissolved particles, not their mass, macromolecules (proteins, nucleic acids, polysaccharides) have far less effect on the osmolarity of a solution than would an equal mass of their monomeric components. For example, a gram of a polysaccharide composed of 1,000 glucose units has the same effect on osmolarity as a milligram of glucose. One effect of storing fuel as polysaccharides (starch or glycogen) rather than as glucose or other simple sugars is prevention of an enormous increase in osmotic pressure within the storage cell.

Plants use osmotic pressure to achieve mechanical rigidity. The very high solute concentration in the plant cell vacuole draws water into the cell (Fig.12). The resulting osmotic pressure against the cell wall (turgor pressure) stiffens the cell, the tissue, and the plant body. When the lettuce in your salad wilts, it is because loss of water has reduced turgor pressure. Sudden alterations in turgor pressure produce the movement of plant parts seen in touch-sensitive plants such as the Venus flytrap and mimosa.

Osmosis also has consequences for laboratory protocols. Mitochondria, chloroplasts, and lysosomes, for example, are bounded by semipermeable membranes. In isolating these organelles from broken cells, biochemists must perform the fractionations in isotonic solutions. Buffers used in cellular fractionations commonly contain sufficient concentrations (about 0.2 M) of sucrose or some other inert solute to protect the organelles from osmotic lysis.

Ionization of Water, Weak Acids,and Weak Bases

Although many of the solvent properties of water can be explained in terms of the uncharged H_2O molecule, the small degree of ionization of water to hydrogen ions (H^+) and hydroxide ions (OH^-) must also be taken into account. Like all reversible reactions, the ionization of water can be described by an equilibrium constant. When weak acids are dissolved in water, they contribute H^+ by ionizing; weak bases consume H^+ by becoming protonated. These processes are also governed by equilibrium constants. The total hydrogen ion concentration from all sources is experimentally measurable and is expressed as the pH of the solution. To predict the state of ionization of solutes in water, we must take into account the relevant equilibrium constants for each ionization reaction.

Pure Water Is Slightly Ionized

Water molecules have a slight tendency to undergo reversible ionization to yield a hydrogen ion (a proton) and a hydroxide ion, giving the equilibrium.

$$H_2O \rightleftharpoons H^+ + OH^- \qquad (2)$$

Although commonly show the dissociation product of water as H^+, free protons do not exist in solution; hydrogen ions formed in water are immediately hydrated to hydronium ions (H_3O^+). Hydrogen bonding between water molecules makes the hydration of dissociating protons virtually instantaneous:

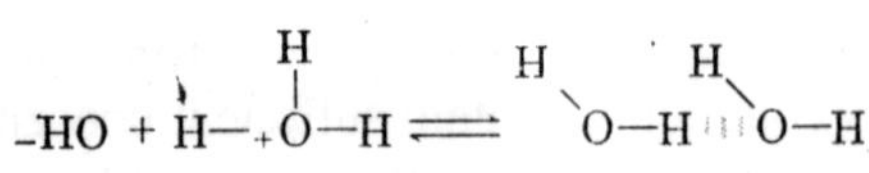

The ionization of water can be measured by its electrical conductivity; pure water carries electrical current as H^+ migrates toward the cathode and OH^- toward the anode. The movement of hydronium and hydroxide ions in the electric field is anomalously fast compared with that of other ions such as Na^+, K^+, and Cl^-. This high ionic mobility results from the kind of "proton hopping" shown in Figure 13. No individual proton moves very far through the bulk solution, but a series of proton hops between hydrogen-bonded water molecules causes the net movement of a proton over a long distance in a remarkably short time. As a result of the high ionic mobility of H^+ (and of OH^- which also moves rapidly by proton hopping, but in the opposite direction), acid-base reactions in aqueous solutions are generally exceptionally fast. As noted above, proton hopping very likely also plays a role in biological proton-transfer reactions.

Because reversible ionization is crucial to the role of water in cellular function, we must have a means of expressing the extent of ionization of water in quantitative terms. A brief review of some properties of reversible chemical reactions shows how this can be done. The position of equilibrium of any chemical reaction is given by its equilibrium constant, K_{eq} (sometimes expressed simply as K). For the generalized reaction

$$A + B \rightleftharpoons C + D \quad (3)$$

an equilibrium constant can be defined in terms of the concentrations of reactants (A and B) and products (C and D) at equilibrium:

$$K_{eq} = \frac{[C][D]}{[A][B]}$$

Strictly speaking, the concentration terms should be the activities, or effective concentrations in nonideal solutions, of each species. Except in very accurate work, however, the equilibrium constant may be approxizy mated by measuring the concentrations at equilibrium. For reasons beyond the scope of this discussion, equilibrium constants are dimensionless.

The equilibrium constant is fixed and characteristic for any given chemical reaction at a specified temperature. It defines the composition of the final equilibrium mixture, regardless of the starting amounts of reactants and products. Conversely, we can calculate the equilibrium constant for a given reaction at a given temperature if the equilibrium concentrations of all its reactants and products are known. The standard free-energy change ($\Delta G°$) is directly related to K_{eq}..

The Ionization of Water Is Expressed by an Equilibrium Constant

The degree of ionization of water at equilibrium (Eqn.2) is small; at 25 °C only about two of every 109 molecules in pure water are ionized at any instant. The equilibrium constant for the reversible ionization of water (Eqn. 2) is

$$K_{eq} = \frac{[H^+][OH^-]}{[H_2O]} \tag{4}$$

In pure water at 25°C, the concentration of water is 55.5 M (grams of H_2O in 1 L divided by its gram molecular weight: (1,000 g/L)/(18.015 g/mol)) and is essentially constant in relation to the very low concentrations of H^+ and OH^-,namely,1×10^{-7}M. Accordingly, we can substitute 55.5 M in the equilibrium constant expression (Eqn. 4) to yield

$$K_{eq} = \frac{[H^+][OH^-]}{55.5M}$$

which, on rearranging, becomes

$$(55.5\ M)(K_{eq}) = [H^+][OH^-] = K_w \tag{5}$$

where K_w designates the product (55.5 M)(K_{eq}), the ion product of water at 25 °C. The value for K_{eq}, determined by electrical-conductivity measurements of pure water, is 1.8×10^{-16} M at 25°C. Substituting this value for K_{eq} in Equation 5 gives the value of the ion product of water:

$$K_w = [H^+][OH^-] = (55.5\ M)(1.8{\times}10^{-16}\ M)$$

$$= 1.0{\times}\ 10^{-14}\ M^2$$

Thus the product $[H^-][OH^+]$ in aqueous solutions at 25°C always equals $1 \times 10^{-14} M^2$. When there are exactly equal concentrations of $H^+$and OH^-, as in pure water, the solution is said to be at neutral pH. At this pH, the concentration of $H^+$and OH^- can be calculated from the ion product of water as follows:

$$K_w = [H^+[OH\ ] = [H^+]^2$$

Solving for $[H^+]$ gives

$$[H^+] = \sqrt{K_w} = \sqrt{1{\times}10^{-4} M^2}\ [H^+] = [OH^-] = 10^{-7}$$

As the ion product of water is constant, whenever $[H^+]$ is greater than $1\ {\times}10^{-7}M$, $[OH^-]$ must become less than $1\ {\times}10^{-7}M$, and vice versa. When $[H^+]$ is very high, as in a solution of hydrochloric acid, $[OH^-]$ must be very low. From the ion product of water we can calculate $[H^+]$ if we know $[OH^-]$. and vice versa.

The pH Scale Designates the H^+ and OH^- Concentrations

The ion product of water, K_w, is the basis for the pH scale. It is a convenient means of designating the concentration of H^+ (and thus of OH^-) in any aqueous solution in the range between 1.0 *M* H^+ and 1.0 M OH^-. The term pH is defined by the expression

$$\text{pH} = \log \frac{1}{[H^+]} = -\log [H^+]$$

The symbol p denotes "negative logarithm of." For a precisely neutral solution at 25°C, in which the concentrationof hydrogen ions is 1.0×10^{-7}M, the pH can be calculated as follows:

$$pH = \log \frac{1}{1.0 \times 10^{-7}} = \log (1.0 \times 10^{7})$$

$$= \log 1.0 + \log 10^{7} = 0 + 7 = 7$$

The value of 7 for the pH of a precisely neutral solution is not an arbitrarily chosen figure; it is derived from the absolute value of the ion product of water at 25°C, which by convenient coincidence is a round number. Solutions having a pH greater than 7 are alkaline or basic; the concentration of OH^- is greater than that of H^+. Conversely, solutions having a pH less than 7 are acidic.

Table 4: The pH scale

$[H^+]$ (M)	pH	$[OH^-]$ (M)	pOH
10^{0} (1)	0	10^{-14}	14
10^{-1}	1	10^{-13}	13
10^{-2}	2	10^{-12}	12
10^{-3}	3	10^{-11}	11
10^{-4}	4	10^{-10}	10
10^{-5}	5	10^{-9}	9
10^{-6}	6	10^{-8}	8
10^{-7}	7	10^{-7}	7
10^{-8}	8	10^{-6}	6
10^{-9}	9	10^{-5}	5
10^{-10}	10	10^{-4}	4
10^{-11}	11	10^{-3}	3
10^{-12}	12	10^{-2}	2
10^{-13}	13	10^{-1}	1
10^{-14}	14	10^{0}(1)	0

Note that the pH scale is logarithmic, not arithmetic. To say that two solutions differ in pH by 1 pH unit means that one solution has ten times the H^+ concentration of the other, but it does not tell us the absolute magnitude of the difference. Figure 14 gives the pH of some common aqueous fluids. A cola drink (pH 3.0) or red wine (pH 3.7) has an H^+ concentration approximately 10,000 times that of blood (pH 7.4).

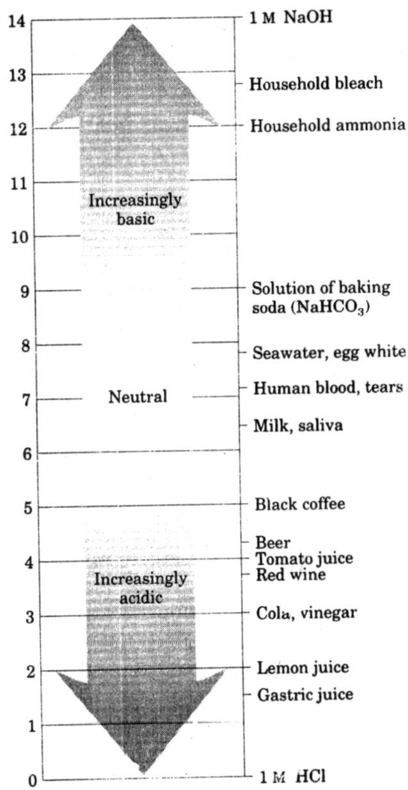

Figure 14: The pH of some aqueous fluids

The pH of an aqueous solution can be approximately measured using various indicator dyes, including litmus, phenolphthalein, and phenol red, which undergo color changes as a proton dissociates from the dye molecule. Accurate determinations of pH in the chemical or clinical laboratory are made with a glass electrode that is selectively sensitive to H^+ concentration but insensitive to Na^+, K^+, and other cations. In a pH meter the signal from such an electrode is amplified and compared with the signal generated by a solution of accurately known pH.

Measurement of pH is one of the most important and frequently used procedures in biochemistry. The pH affects the structure and activity of biological macromolecules; for example, the catalytic activity of enzymes is strongly dependent on pH. Measurements of the pH of blood and urine are commonly used in medical diagnoses. The pH of the blood plasma of people with severe, uncontrolled diabetes, for example, is often below the normal value of 7.4; this condition is called acidosis. In certain other disease states the pH of the blood is higher than normal, the condition of alkalosis.

Hydrochloric, sulfuric, and nitric acids, commonly called strong acids, are completely ionized in dilute aqueous solutions; the strong bases NaOH and KOH are also completely ionized. Of more interest to biochemists is the behavior of weak acids and bases-those not completely ionized when dissolved in water. These are common in biological systems and play important roles in metabolism and its regulation.

The behavior of aqueous solutions of weak acids and bases is best understood if we first define some terms.Acids may be defined as proton donors and bases as proton acceptors. A proton donor and its corresponding proton acceptor make up a conjugate acid-base pair (Fig.15). Acetic acid (CH_3COOH), a proton donor, and the acetate anion (CH_3COO^-), the corresponding proton acceptor, constitute a conjugate acidbase pair, related by the reversible reaction

$$CH_3COOH \rightleftharpoons H^+ + CH_3COO$$

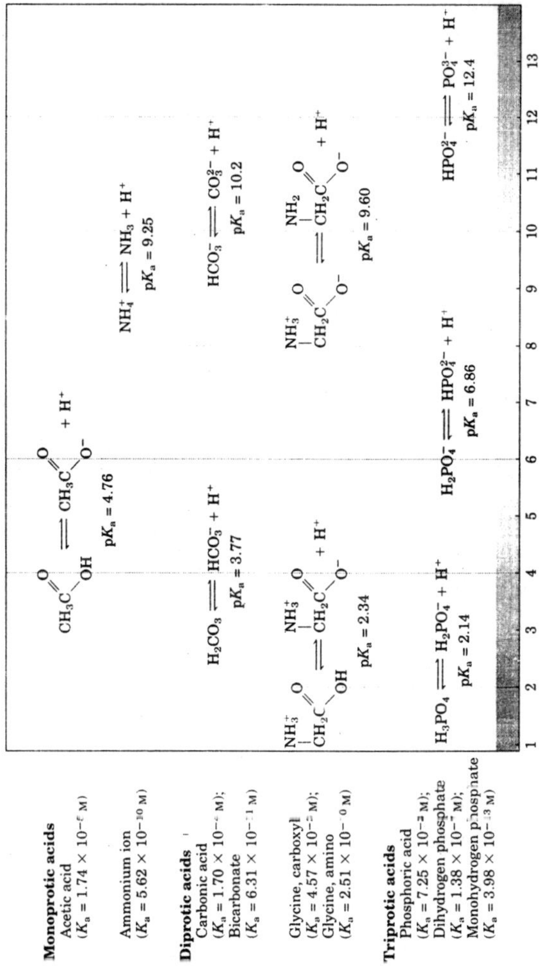

Figure 15: Conjugate acid-base pairs consist of a proton donor and a proton acceptor

Each acid has a characteristic tendency to lose its proton in an aqueous solution. The stronger the acid, the greater its tendency to lose its proton. The tendency of any acid (HA) to lose a proton and form its conjugate base (A^-) is defined by the equilibrium constant (K_{eq}) for the reversible reaction

$$HA \rightleftharpoons H^+ + A^-,$$

which is

$$K_{eq} = \frac{[H^+][A^-]}{[HA]} = K_a$$

Equilibrium constants for ionization reactions are usually called ionization or dissociation constants, often designated K_a. The dissociation constants of some acids are given in Figure15. Stronger acids, such as phosphoric and carbonic acids, have larger dissociation constants; weaker acids, such as monohydrogen phosphate(HPO_4^{2-}), have smaller dissociation constants.

Also included in Figure 15 are values of pKa, which is analogous to pH and is defined by the equation

$$pK_a = \log \frac{1}{K_a} = -\log K_a$$

The stronger the tendency to dissociate a proton, the stronger is the acid and the lower its pK_a.

Titration Curves

Titration is used to determine the amount of an acid in a given solution. A measured volume of the acid is titrated with a solution of a strong base, usually sodium hydroxide (NaOH), of known concentration. The NaOH is added in small increments until the acid is consumed (neutralized), as determined with an indicator dye or a pH meter. The concentration of the acid in the original solution can be calculated from the volume and concentration of NaOH added.

A plot of pH against the amount of NaOH added (a titration curve) reveals the pKa of the weak acid. Consider the titration of a 0.1 M solution of acetic acid (for simplicity denoted as HAc) with 0.1 M NaOH at 25°C (Fig. 16). Two reversible equilibria are involved in the process:

$$H_2O \rightleftharpoons H^+ + OH^- \qquad (6)$$

$$HAc \rightleftharpoons H^+ + Ac^- \qquad (7)$$

The equilibria must simultaneously conform to their characteristic equilibrium constants, which are, respectively,

$$K_w = [H^+][OH^-] = 1 \times 10^{-14}\ \text{M}^2 \qquad (8)$$

$$K_a = \frac{[H^+][Ac^-]}{[HAc]} = 1.74 \times 10^5\ \text{M} \qquad (9)$$

At the beginning of the titration, before any NaOH is added, the acetic acid is already slightly ionized, to an extent that can be calculated from its dissociation constant (Eqn 9).

As NaOH is gradually introduced, the added OH^- combines with the free H^-in the solution to form H2O, to an extent that satisfies the equilibrium relationship in Equation 8. As free H^+ is removed, HAc dissociates further to satisfy its own equilibrium constant (Eqn 9). The net result as the titration proceeds is that more and more HAc ionizes; forming Ac^-, as the NaOH is added. At the midpoint of the titration, at which exactly 0.5 equivalent of NaOH has been added, one-half of the original acetic acid has undergone dissociation, so that the concentration of the proton donor, [HAc], now equals that of the proton acceptor, $[Ac^-]$. At this midpoint a very important relationship holds: the pH of the equimolar solution of acetic acid and acetate is exactly equal to the pK_a of acetic acid (pK_a = 4.76; Figs 15 ,16). The basis for this relationship, which holds for all weak acids, will soon become clear.

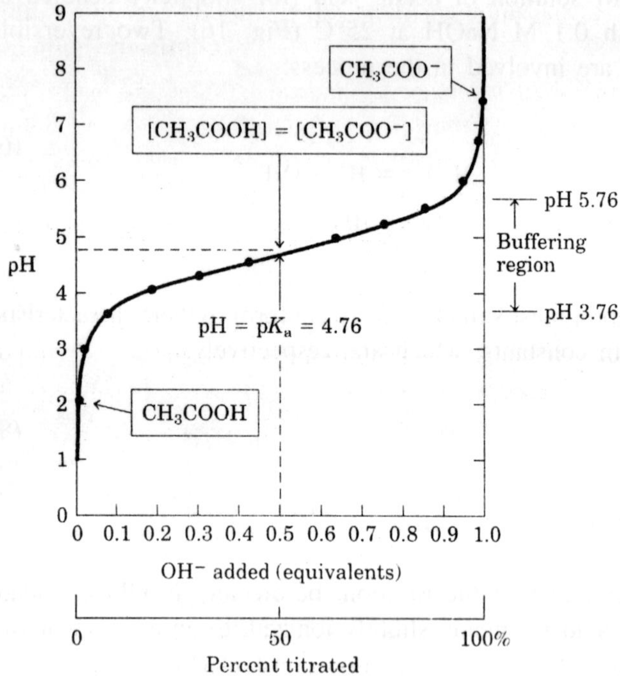

Figure 16: The titration curve of acetic acid

As the titration is continued by adding further increments of NaOH, the remaining nondissociated acetic acid is gradually converted into acetate. The end point of the titration occurs at about pH 7.0: all the acetic acid has lost its protons to OH^-, to form H_2O and acetate. Throughout the titration the two equilibria (Eqns 6, 7) coexist, each always conforming to its equilibrium constant.

Figure 17 compares the titration curves of three weak acids with very different dissociation constants: acetic acid (pK_a=4.76);

dihydrogen phosphate, H_2PO_4 (pK_a = 6.86); and ammonium ion, NH^+_4 = (pK_a = 9.25).

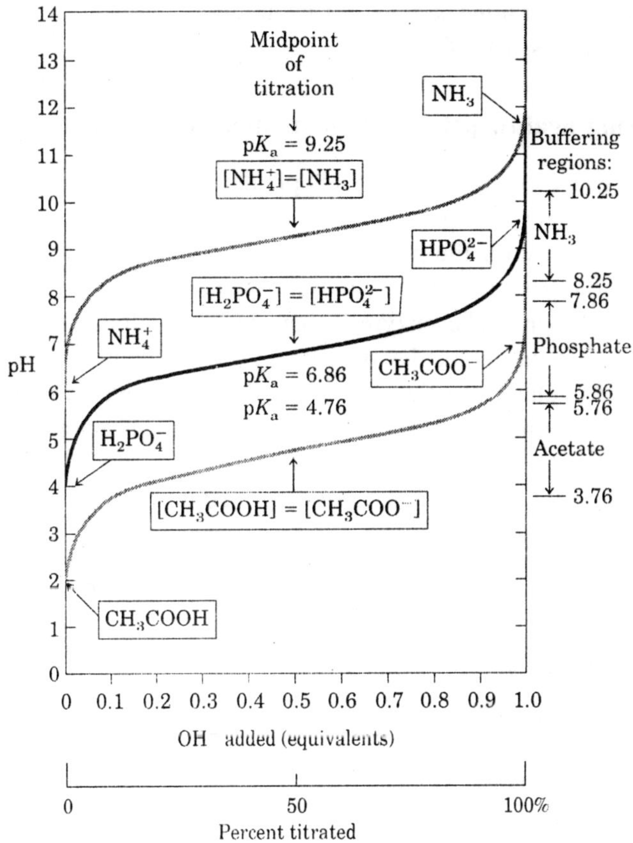

Figure 17: Comparison of the titration curves of three weak acids

Although the titration curves of these acids have the same shape, they are displaced along the pH axis because the three acids have different strengths. Acetic acid, with the highest Ka (lowest pKa) of the three, is the strongest (loses its proton most readily); it is already half dissociated at pH 4.76. Dihydrogen phosphate loses a proton less readily, being half dissociated at pH 6.86.

Ammonium ion is the weakest acid of the three and does not become half dissociated until pH 9.25. The most important point about the titration curve of a weak acid is that it shows graphically that a weak acid and its anion-a conjugate acid-base pair-can act as a buffer.

Buffering against pH Changes in Biological Systems

Almost every biological process is pH dependent; a small change in pH produces a large change in the rate of the process. This is true not only for the many reactions in which the H^+ ion is a direct participant, but also for those in which there is no apparent role for H^+ ions. The enzymes that catalyze cellular reactions, and many of the molecules on which they act, contain ionizable groups with characteristic pK_a values. The protonated amino and carboxyl groups of amino acids and the phosphate groups of nucleotides, for example, function as weak acids; their ionic state depends on the pH of the surrounding medium. As noted above, ionic interactions are among the forces that stabilize a protein molecule and allow an enzyme to recognize and bind its substrate. Cells and organisms maintain a specific and constant cytosolic pH, keeping biomolecules in their optimal ionic state, usually near pH 7. In multicellular organisms, the pH of extracellular fluids is also tightly regulated. Constancy of pH is achieved primarily by biological buffers: mixtures of weak acids and their conjugate bases.

Buffers Are Mixtures of Weak Acids and Their Conjugate Bases

Buffers are aqueous systems that tend to resist changes in pH when small amounts of acid (H^+) or base (OH^-) are added. A buffer system consists of a weak acid (the proton donor) and its conjugate base (the proton acceptor). As an example, a mixture of equal concentrations of acetic acid and acetate ion, found at the midpoint of the titration curve in Figure 16, is a buffer system. The titration curve of acetic acid has a relatively flat zone extending about 1 pH unit on either side of its midpoint pH of 4.76. In this zone, an amount of H^+ or OH^- added to the system has much less effect on pH than the same amount added outside the buffer range. This relatively flat zone is the buffering region of the acetic acid-acetate buffer pair. At the midpoint of the buffering region, where the concentration of the proton donor (acetic acid) exactly equals that of the proton acceptor (acetate), the buffering power of the system is maximal; that is, its pH changes least on addition of H^+ or OH^-. The pH at this point in the titration curve of acetic acid is equal to its pKa. The pH of the acetate buffer system does change slightly when a small amount of H^+ or OH^- is added, but this change is very small compared with the pH change that would result if the same amount of H^+ or OH^- were added to pure water or to a solution of the salt of a strong acid and strong base, such as NaCl, which has no buffering power.

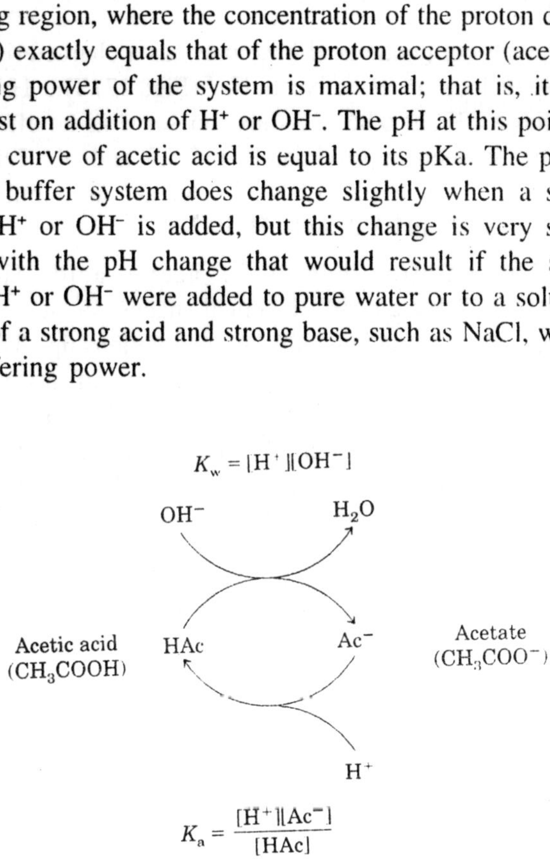

Figure 18: The acetic acid-acetate pair as a buffer system

Buffering results from two reversible reaction equilibria occurring in a solution of nearly equal concentrations of a proton donor and its conjugate proton acceptor. Figure 18 explains how a buffer system works. Whenever H^+or OH^- is added to a buffer, the result is a small change in the ratio of the relative concentrations of the weak acid and its anion and thus a small change in pH. The decrease in concentration of one component of the system is balanced exactly by an increase in the other. The sum of the buffer components does not change, only their ratio.

Each conjugate acid-base pair has a characteristic pH zone in which it is an effective buffer (Fig. 17).The $H_2PO_4^-/HPO_4^{2-}$ pair has a pK_a of 6.86 and thus can serve as an effective buffer system between approximately pH 5.9 and pH 7.9; the NH_4^+/NH_3 pair, with a pK_a of 9.25, can act as a buffer between approximately pH 8.3 and pH 10.3.

The titration curves of acetic acid, $H_2PO_4^-$, and NH_4^+(Fig.17) have nearly identical shapes, suggesting that these curves reflect a fundamental law or relationship. This is indeed the case. The shape of the titration curve of any weak acid is described by the Henderson- Hasselbalch equation, which is important for understanding buffer action and acid-base balance in the blood and tissues of vertebrates. This equation is simply a useful way of restating the expression for the dissociation constant of an acid. For the dissociation of a weak acid HA into H^+ and A^-, the Henderson-Hasselbalch equation can be derived as follows:

$$K_a = \frac{[H^+][A^-]}{[HA]}$$

First solve for $[H^+]$

$$[H^+] = K_a\frac{[HA]}{[A^-]}$$

Then take the negative logarithm of both sides:

$$-\log [H^+] = -\log K_a - \log \frac{[HA]}{[A^-]}$$

Substitute pH for –log [H] and pK_a for –log K_a:

$$pH = pK_a - \log \frac{[HA]}{[A^-]}$$

Now invert –log [HA]/[A⁻], which involves changing its sign, to obtain the Henderson-Hasselbalch equation:

$$pH = pK_a + \log \frac{[A^-]}{[HA]} \qquad (10)$$

Stated more generally,

$$pH = pK_a + \log \frac{[\text{proton acceptor}]}{[\text{proton donor}]}$$

This equation fits the titration curve of all weak acids and enables us to deduce a number of important quantitative relationships. For example, it shows why the pK_a of a weak acid is equal to the pH of the solution at the midpoint of its titration. At that point, [HA] equals [A^-],and

$$pH = pK_a + \log 1 = pK_a + 0 = pK_a$$

The intracellular and extracellular fluids of multicellular organisms have a characteristic and nearly constant pH. The organism's first line of defense against changes in internal pH is provided by buffer systems. The cytoplasm of most cells contains high concentrations of proteins, which contain many amino acids with functional groups that are weak acids or weak bases. For example, the side chain of histidine (Fig. 19) has a pK_a of 6.0; proteins containing histidine residues therefore buffer effectively near neutral pH. Nucleotides such as ATP, as well as many low molecular weight metabolites, contain ionizable groups that can contribute buffering power to the cytoplasm.

Some highly specialized organelles and extracellular compartments have high concentrations of compounds that contribute buffering capacity: organic acids buffer the vacuoles of plant cells; ammonia buffers urine.

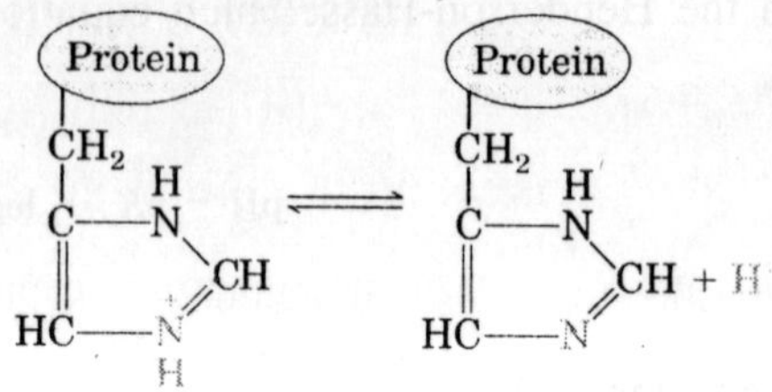

Figure 19: The amino acid histidine, a component of protiens. is a weak acid. The pK_a *of the protonated nitrogen of the side chain is 6.0*

Two especially important biological buffers are the phosphate and bicarbonate systems. The phosphate buffer system, which acts in the cytoplasm of all cells, consists of $H_2PO_4^-$ as proton donor and HPO_4^{2-} as proton acceptor:

$$H_2PO_4^- \rightleftharpoons H^+ + HPO_4^{2-}$$

The phosphate buffer system is maximally effective at a pH close to its pK_a of 6.86 (Figs 15, 17) and thus tends to resist pH changes in the range between about 5.9 and 7.9. It is therefore an effective buffer in biological fluids; in mammals, for example, extracellular fluids and most cytoplasmic compartments have a pH in the range of 6.9 to 7.4.Blood plasma is buffered in part by the bicarbonate system, consisting of carbonic acid (H_2CO_3) as proton donor and bicarbonate (HCO_3^-) as proton acceptor:

$$H_2CO_3 \rightleftharpoons H^+ + HCO_3^-$$

$$K_1 = \frac{[H^+][HCO_3^-]}{[H_2CO_3]}$$

This buffer system is more complex than other conjugate acid-base pairs because one of its components, carbonic acid (H_2CO_3), is formed from dissolved (d) carbon dioxide and water, in a reversible reaction:

$$CO_2(d) + H_2O \rightleftharpoons H_2CO_3$$

$$K_2 = \frac{[H_2CO_3]}{[CO_2(d)][H_2O]}$$

Carbon dioxide is a gas under normal conditions, and the concentration of dissolved CO_2 is the result of equilibration with CO2 of the gas (g) phase:

$$CO_2(g) \rightleftharpoons CO_2(d)$$

$$K_3 = \frac{[CO_2(d)]}{[CO_2(g)]}$$

The pH of a bicarbonate buffer system depends on the concentration of H_2CO_3 and HCO_3^-, the proton donor and acceptor components. The concentration of H_2CO_3 in turn depends on the concentration of dissolved CO_2, which in turn depends on the concentration of CO_2 in the gas phase, called the partial pressure of CO_2. Thus the pH of a bicarbonate buffer exposed to a gas phase is ultimately determined by the concentration of HCO_3^-in the aqueous phase and the partial pressure of CO_2 in the gas phase. Human blood plasma normally has a pH close to 7.4. Should the pH-regulating mechanisms fail or be overwhelmed, as may happen in severe uncontrolled diabetes when an overproduction of metabolic acids causes acidosis, the pH of the blood can fall to 6.8 or below, leading to irreparable cell damage and death. In other diseases the pH may rise to lethal levels.

Although many aspects of cell structure and function are influenced by pH, it is the catalytic activity of enzymes that is especially sensitive. Enzymes typically show maximal catalytic activity at a characteristic pH, called the pH optimum (Fig.20). On either side of the optimum pH their catalytic activity often declines sharply. Thus, a small change in pH can make a large difference in the rate of some crucial enzyme-catalyzed reactions. Biological control of the pH of cells and body fluids is therefore of central importance in all aspects of metabolism and cellular activities.

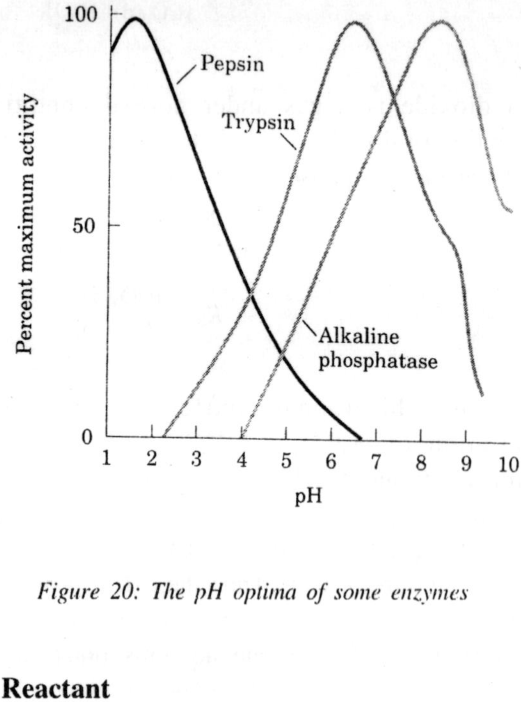

Figure 20: The pH optima of some enzymes

Water as a Reactant

Water is not just the solvent in which the chemical reactions of living cells occur; it is very often a direct participant in those reactions. The formation of ATP from ADP and inorganic phosphate is an example of a condensation reaction in which the elements of water are eliminated (Fig. 21a). The reverse of this reaction-cleavage accompanied by the addition of the elements of water-is a hydrolysis reaction. Hydrolysis reactions are also responsible for the enzymatic depolymerization of proteins, carbohydrates, and nucleic acids.

Hydrolysis reactions, catalyzed by enzymes called hydrolases, are almost invariably exergonic. The formation of cellular polymers from their subunits by simple reversal of hydrolysis would be endergonic and therefore does not occur.

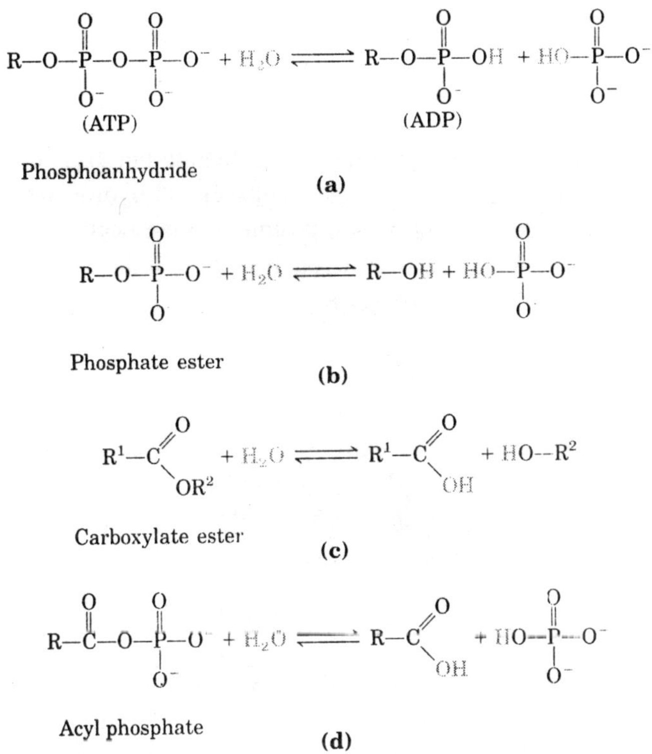

Figure 21: Participation of water in biological reaction

Water and carbon dioxide are the end products of the oxidation of fuels such as glucose. The overall reaction can be summarized as

The "metabolic water" formed by oxidation of foods and stored fats is actually enough to allow some animals in very dry habitats (gerbils, kangaroo rats, camels) to survive for extended periods without drinking water.

The CO_2 produced by glucose oxidation is converted in erythrocytes to the more soluble HCO_3^-, in a reaction catalyzed by the enzyme carbonic anhydrase:

$$CO_2 + H_2O \rightleftharpoons HCO_3^- + H^+$$

In this reaction, water not only is a substrate but also functions in proton transfer by forming a network of hydrogen-bonded water molecules through which proton hopping occurs (Fig. 13). Green plants and algae use the energy of sunlight to split water in the process of photosynthesis:

$$2H_2O + 2A \xrightarrow{\text{light}} O_2 + 2AH_2$$

In this reaction, A is an electron-accepting species, which varies with the type of photosynthetic organism, and water serves as the electron donor in an oxidationreduction sequence that is fundamental to all life.

The Fitness of the AqueousEnvironment

Organisms have effectively adapted to their aqueous environment and have evolved means of exploiting the unusual properties of water. The high specific heat of water (the heat energy required to raise the temperature of 1 g of water by 1°C) is useful to cells and orzyganisms because it allows water to act as a "heat buffer," keeping the temperature of an organism relatively constant as the temperature of the surroundings fluctuates and as heat is generated as a byproduct of metabolism. Furthermore, some vertebrates exploit the high heat of vaporization of water by using (thus losing) excess body heat to evaporate sweat.

The high degree of internal cohesion of liquid water, due to hydrogen bonding, is exploited by plants as a means of transporting dissolved nutrients from the roots to the leaves during the process of transpiration. Even the density of ice, lower than that of liquid water, has important biological consequences in the life cycles of aquatic organisms. Ponds freeze from the top down, and the layer of ice at the top insulates the water below from frigid air, preventing the pond (and the organisms in it) from freezing solid. Most fundamental to all living organisms is the fact that many physical and biological properties of cell macromolecules, particularly the proteins and nucleic acids, derive from their interactions with water molecules of the surrounding medium. The influence of water on the course of biological evolution has been profound and determinative. If life forms have evolved elsewhere in the universe, they are unlikely to resemble those of Earth unless their extraterrestrial origin is also a place in which plentiful liquid water is available.

3

Amino Acids

Proteins are the most abundant biological macromolecules, occurring in all cells and all parts of cells. Proteins also occur in great variety; thousands of different kinds, ranging in size from relatively small peptides to huge polymers with molecular weights in the millions, may be found in a single cell. Moreover, proteins exhibit enormous diversity of biological function.

Proteins are the molecular instruments through which genetic information is expressed. Relatively simple monomeric subunits provide the key to the structure of the thousands of different proteins. All proteins, whether from the most ancient lines of bacteria or from the most complex forms of life, are constructed from the same ubiquitous set of 20 amino acids, covalently linked in characteristic linear sequences.

Because each of these amino acids has a side chain with distinctive chemical properties, this group of 20 precursor molecules may be regarded as the alphabet in which the language of protein structure is written. What is most remarkable is that cells can produce proteins with strikingly different properties and activities by joining the same 20 amino acids in many different combinations and sequences.

From these building blocks different organisms can make such widely diverse products as enzymes, hormones, antibodies, transporters, muscle fibers, the lens protein of the eye, feathers, spider webs, rhinoceros horn, milk proteins, antibiotics, mushroom poisons, and myriad other substances having distinct biological activities. Among these protein products, the enzymes are the most varied and specialized. Virtually all cellular reactions are catalyzed by enzymes.

General Structure of Amino Acids

Proteins are polymers of amino acids, with each amino acid residue joined to its neighbor by a specific type of covalent bond. (The term "residue" reflects the loss of the elements of water when one amino acid is joined to another.) Proteins can be broken down (hydrolyzed) to their constituent amino acids by a variety of methods, and the earliest studies of proteins naturally focused on the free amino acids derived from them. Twenty different amino acids are commonly found in proteins. The first to be discovered was asparagine, in 1806. The last of the 20 to be found, threonine, was not identified until 1938.

All the amino acids have trivial or common names, in some cases derived from the source from which they were first isolated. Asparagine was first found in asparagus, and glutamate in wheat gluten; tyrosine was first isolated from cheese (its name is derived from the Greek tyros, "cheese"); and glycine (Greek glykos, "sweet") was so named because of its sweet taste.

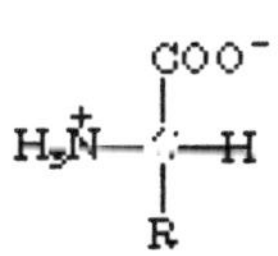

Figure.1.General structure of an amino acid

All 20 of the common amino acids are α-amino acids. They have a carboxyl group and an amino group bonded to the same carbon atom (the α carbon) (Figure 1). They differ from each other in their side chains, or R groups, which vary in structure, size, and electric charge, and which influence the solubility of the amino acids in water. In addition to these 20 amino acids there are many less common ones. Some are residues modified after a protein has been synthesized; others are amino acids present in living organisms but not as constituents of proteins.

The common amino acids of proteins have been assigned three-letter abbreviations and one-letter symbols (Table 1), which are used as shorthand to indicate the composition and sequence of amino acids polymerized in proteins. Two conventions are used to identify the carbons in an amino acid-a practice that can be confusing. The additional carbons in an R group are commonly designated β γ δ ε and so forth, proceeding out from the α carbon.

For most other organic molecules, carbon atoms are simply numbered from one end, giving highest priority (C-1) to the carbon with the substituent containing the atom of highest atomic number. Within this latter convention, the carboxyl carbon of an amino acid would be C-1 and the α carbon would be C-2. In some cases, such as amino acids with heterocyclic R groups, the Greek lettering system is ambiguous and the numbering convention is therefore used.

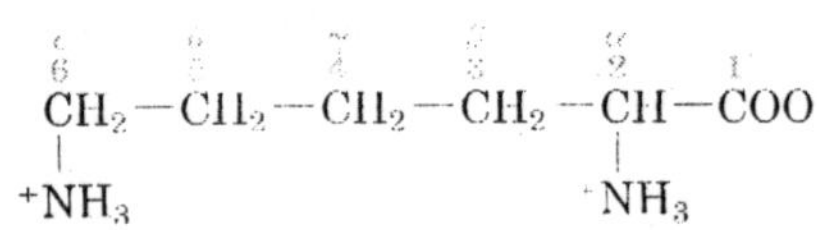

Lysine

For all the common amino acids except glycine, the α carbon is bonded to four different groups: a carboxyl group, an amino group, an R group, and a hydrogen atom (Figure 1; in glycine, the R group is another hydrogen atom). The α-carbon atom is thus a chiral center. Because of the tetrahedral arrangement of the bonding orbitals around the α-carbon atom, the four different groups can occupy two unique spatial arrangements, and thus amino acids have two possible stereoisomers.

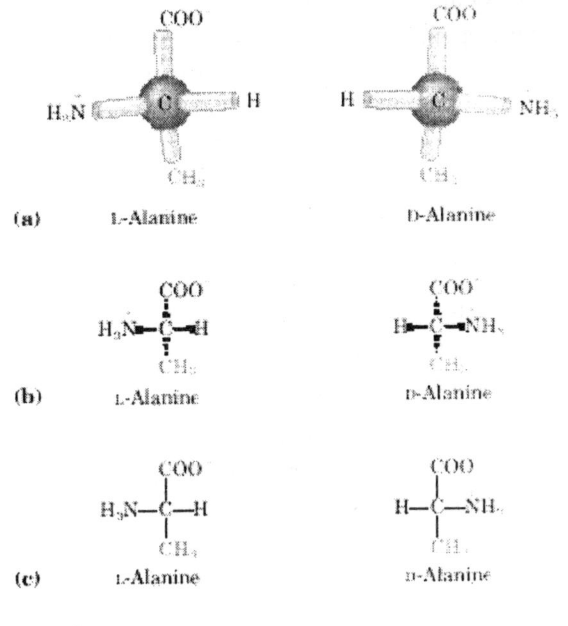

Figure 2: Stereoisomerism in α amino acids.

Since they are nonsuperimposable mirror images of each other (Figure 2), the two forms represent a class of stereoisomers called enantiomers. (a) The two stereoisomers of alanine, *L*- and D alanine, are nonsuperimposable mirror images of each other (enantiomers). (b, c) Two different conventions for showing the configurations in space of stereoisomers. In perspective formulas (b) the solid wedge-shaped bonds project out of the plane of the paper, the dashed bonds behind it. In projection formulas

(c) the horizontal bonds are assumed to project out of the plane of the paper, the vertical bonds behind. However, projection formulas are often used casually and are not always intended to portray a specific stereochemical configuration.

All molecules with a chiral center are also optically active-that is, they rotate plane-polarized light. Special nomenclature has been developed to specify the absolute configuration of the four substituents of asymmetric carbon atoms. The absolute configurations of simple sugars and amino acids are specified by the *D, L system* (Fig. 3), based on the absolute configuration of the three-carbon sugar glyceraldehyde, a convention proposed by Emil Fischer in 1891. For all chiral compounds, stereoisomers having a configuration related to that of *L*-glyceraldehyde are designated *L*, and stereoisomers related to D-glyceraldehyde are designated D. The functional groups of L-alanine are matched with those of Lglyceraldehyde by aligning those that can be interconverted by simple, one-step chemical reactions.

Thus the carboxyl group of L-alanine occupies the same position about the chiral carbon as does the aldehyde group of L-glyceraldehyde, because an aldehyde is readily converted to a carboxyl group via a one-step oxidation. Historically, the similar *l* and *d* designations were used for levorotatory (rotating light to the left) and dextrorotatory (rotating light to the right).

However, not all (blue) is different in each amino acid. *L*-amino acids are levorotatory, and the convention shown in Figure 3 was needed to avoid potential ambiguities about absolute configuration. By Fischer's convention, *L* and *D* refer only to the absolute configuration of the four substituents around the chiral carbon, not to optical properties of the molecule. Another system of specifying configuration around a chiral center is the RS system, which is used in the systematic nomenclature of organic chemistry and describes more precisely the configuration of molecules with more than one chiral center.

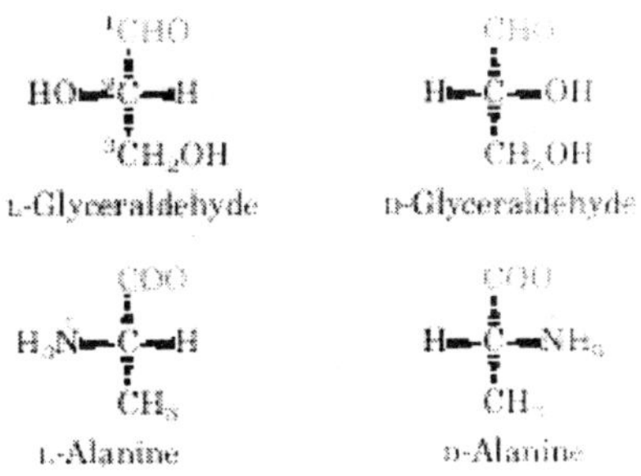

Figure 3: Steric relationship of the stereoisomers of alanine to the absolute configuration of L- and D-glyceraldehyde.

L Stereoisomers

Nearly all biological compounds with a chiral center occur naturally in only one stereoisomeric form, either *D* or *L*. The amino acid residues in protein molecules are exclusively L stereoisomers. *D*-Amino acid residues have been found only in a few, generally small peptides, including some peptides of bacterial cell walls and certain peptide antibiotics.

It is remarkable that virtually all amino acid residues in proteins are *L* stereoisomers. When chiral compounds are formed by ordinary chemical reactions, the result is a racemic mixture of *D* and *L* isomers, which are difficult for a chemist to distinguish and separate. But to a living system, *D* and *L* isomers are as different as the right hand and the left. The formation of stable, repeating substructures in proteins generally requires that their constituent amino acids be of one stereochemical series.

Cells are able to specifically synthesize the L isomers of amino acids because the active sites of enzymes are asymmetric, causing the reactions they catalyze to be stereospecific.

Amino acid	Abbreviation/ symbol	M_r	pK_a values: pK_1 (—COOH)	pK_2 ($—NH_3^+$)	pK_R (R group)	pI	Hydropathy index*	Occurrence in proteins (%)[1]
Nonpolar, aliphatic R groups								
Glycine	Gly G	75	2.34	9.60		5.97	−0.4	7.2
Alanine	Ala A	89	2.34	9.69		6.01	1.8	7.8
Proline	Pro P	115	1.99	10.96		6.48	1.6	5.2
Valine	Val V	117	2.32	9.62		5.97	4.2	6.6
Leucine	Leu L	131	2.36	9.60		5.98	3.8	9.1
Isoleucine	Ile I	131	2.36	9.68		6.02	4.5	5.3
Methionine	Met M	149	2.28	9.21		5.74	1.9	2.3
Aromatic R groups								
Phenylalanine	Phe F	165	1.83	9.13		5.48	2.8	3.9
Tyrosine	Tyr Y	181	2.20	9.11	10.07	5.66	−1.3	3.2
Tryptophan	Trp W	204	2.38	9.39		5.89	−0.9	1.4
Polar, uncharged R groups								
Serine	Ser S	105	2.21	9.15		5.68	−0.8	6.8
Threonine	Thr T	119	2.11	9.62		5.87	−0.7	5.9
Cysteine	Cys C	121	1.96	10.28	8.18	5.07	2.5	1.9
Asparagine	Asn N	132	2.02	8.80		5.41	−3.5	4.3
Glutamine	Gln Q	146	2.17	9.13		5.65	−3.5	4.2
Positively charged R groups								
Lysine	Lys K	146	2.18	8.95	10.53	9.74	−3.9	5.9
Histidine	His H	155	1.82	9.17	6.00	7.59	−3.2	2.3
Arginine	Arg R	174	2.17	9.04	12.48	10.76	−4.5	5.1
Negatively charged R groups								
Aspartate	Asp D	133	1.88	9.60	3.65	2.77	−3.5	5.3
Glutamate	Glu E	147	2.19	9.67	4.25	3.22	−3.5	6.3

*A scale combining hydrophobicity and hydrophilicity of R groups; it can be used to measure the tendency of an amino acid to seek an aqueous environment (− values) or a hydrophobic environment (+ values). See Chapter 11. From Kyte, J. & Doolittle, R.F. (1982) A simple method for displaying the hydropathic character of a protein. *J. Mol. Biol.* **157**, 105–132.

[1]Average occurrence in more than 1,150 proteins. From Doolittle, R.F. (1989) Redundancies in protein sequences. In *Prediction of Protein Structure and the Principles of Protein Conformation* (Fasman, G.D., ed.), pp. 599–623. Plenum Press, New York.

Table 1. Properties and Conventions Associated with the Common Amino Acids Found in Proteins

Classification of Amino Acids

Knowledge of the chemical properties of the common amino acids is central to an understanding of biochemistry. The topic can be simplified by grouping the amino acids into five main classes based on the properties of their R groups (Table 1), in particular, their polarity, or tendency to interact with water at biological pH (near pH 7.0). The polarity of the R groups varies widely, from nonpolar and hydrophobic (water-insoluble) to highly polar and hydrophilic (water-soluble).

The structures of the 20 common amino acids are shown in Figure 4, and some of their properties are listed in Table 1. Within each class there are gradations of polarity, size, and shape of the R groups.

Nonpolar, Aliphatic R Groups

The R groups in this class of amino acids are nonpolar and hydrophobic. The side chains of alanine, valine, leucine, and isoleucine tend to cluster together within proteins, stabilizing protein structure by means of hydrophobic interactions. Glycine has the simplest structure. Although it is formally nonpolar, its very small side chain makes no real contribution to hydrophobic interactions. Methionine, one of the two sulfur-containing amino acids, has a nonpolar thioether group in its side chain.

Proline has an aliphatic side chain with a distinctive cyclic structure. The secondary amino (imino) group of proline residues is held in a rigid conformation that reduces the structural flexibility of polypeptide regions containing proline.

Aromatic R Groups

Phenylalanine, tyrosine, and tryptophan, with their aromatic side chains, are relatively nonpolar (hydrophobic). All can participate in hydrophobic interactions. The hydroxyl group of tyrosine can form hydrogen bonds, and it is an important functional group in

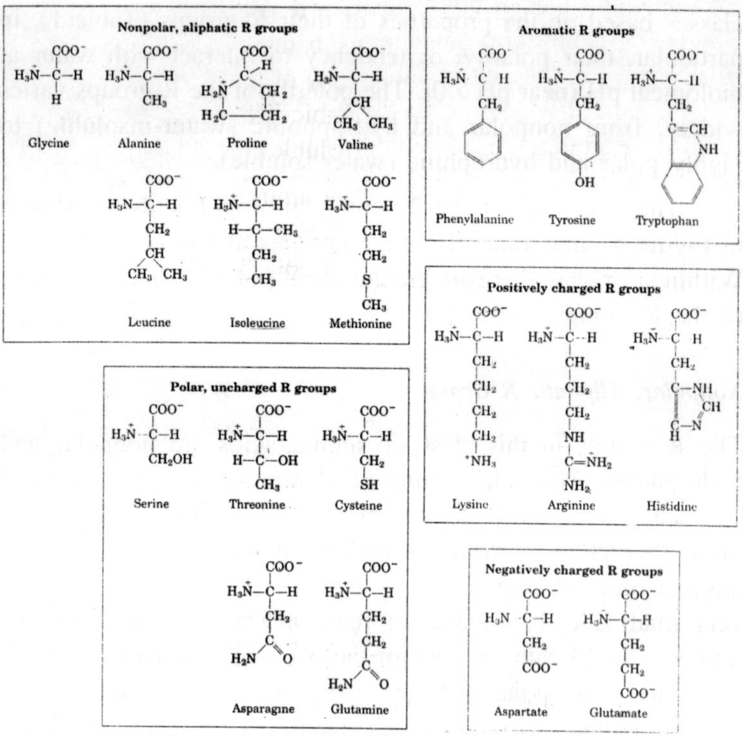

Figure 4: The 20 common amino acids of proteins.

some enzymes. Tyrosine and tryptophan are significantly more polar than phenylalanine, because of the tyrosine hydroxyl group and the nitrogen of the tryptophan indole ring.

Tryptophan and tyrosine, and to a much lesser extent phenylalanine, absorb ultraviolet light (Fig. 5). This accounts for the characteristic strong absorbance of light by most proteins at a wavelength of 280 nm, a property exploited by researchers in the characterization of proteins.

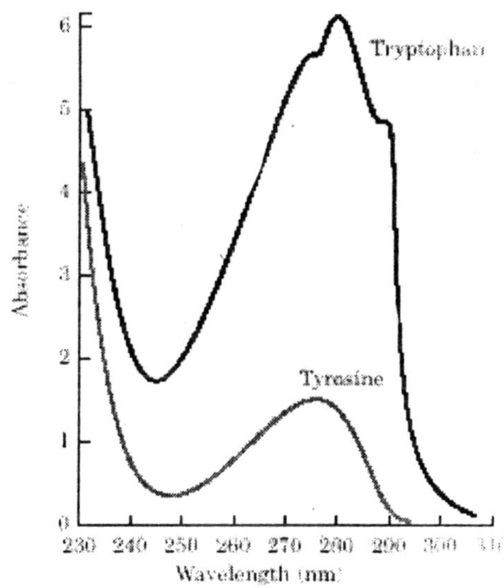

Figure 5: Absorption of ultraviolet light by aromatic amino acids.

Polar, Uncharged R Groups

The R groups of these amino acids are more soluble in water, or more hydrophilic, than those of the nonpolar amino acids, because they contain functional groups that form hydrogen bonds with water. This class of amino acids includes serine, threonine, cysteine, asparagine, and glutamine. The polarity of serine and threonine is contributed by their hydroxyl groups; that of cysteine by its sulfhydryl group; and that of asparagine and glutamine by their amide groups.

Asparagine and glutamine are the amides of two other amino acids also found in proteins, aspartate and glutamate, respectively, to which asparagine and glutamine are easily hydrolyzed by acid or base. Cysteine is readily oxidized to form a covalently linked dimeric amino acid called cystine, in which two cysteine molecules or residues are joined by a disulfide bond (Fig. 6). The disulfide-linked residues are strongly hydrophobic (nonpolar). Disulfide bonds play a special role in the structures of many proteins by forming covalent links between parts of a protein molecule or between two different polypeptide chains.

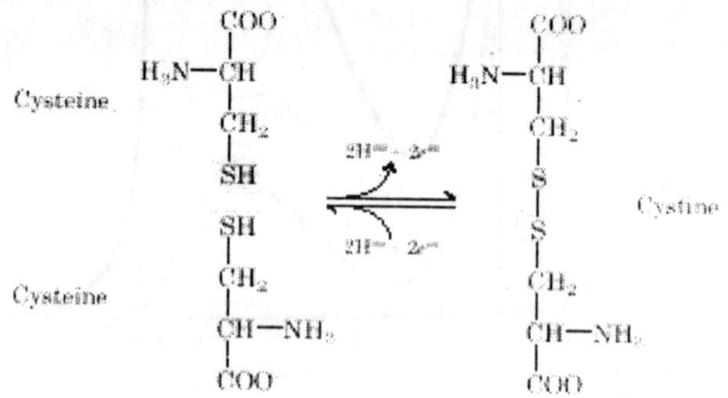

Figure 6: Reversible formation of a disulfide bond by the oxidation of two molecules of cysteine.

Positively Charged (Basic) R Groups

The most hydrophilic R groups are those that are either positively or negatively charged. The amino acids in which the R groups have significant positive charge at pH 7.0 are lysine, which has a second primary amino group at the α position position on its aliphatic chain; arginine, which has a positively charged guanidino group; and histidine, which has an imidazole group. Histidine is the only common amino acid having an ionizable side chain with a pKa near neutrality.

In many enzyme-catalyzed reactions, a His residue facilitates the reaction by serving as a proton donor/acceptor.

Negatively Charged (Acidic) R Groups

The two amino acids having R groups with a net negative charge at pH 7.0 are aspartate and glutamate, each of which has a second carboxyl group.

Functions of Uncommon Amino Acids

In addition to the 20 common amino acids, proteins may contain residues created by modification of common residues already incorporated into a polypeptide (Fig.7). In figure 7, (a) Some uncommon amino acids found in proteins. All are derived from common amino acids. Extra functional groups added by modification reactions are shown in figure 7. Desmosine is formed from four Lys residues (the four carbon back-bones are shaded in yellow). Note the use of either numbers or Greek letters to identify the carbon atoms in these structures. (b) Ornithine and citrulline, which are not found in proteins, are intermediates in the biosynthesis of arginine and in the urea cycle.

Among these uncommon amino acids are 4 hydroxyproline, a derivative of proline, and 5-hydroxylysine, derived from lysine. The former is found in plant cell wall proteins, and both are found in collagen, a fibrous protein of connective tissues. 6-NMethyllysine is a constituent of myosin, a contractile protein of muscle.

Another important uncommon amino acid is γ-carboxyglutamate, found in the bloodclotting protein prothrombin and in certain other proteins that bind Ca^{2+} as part of their biological function. More complex is desmosine, a derivative of

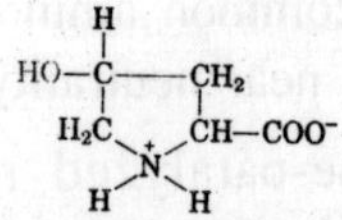

4-Hydroxyproline

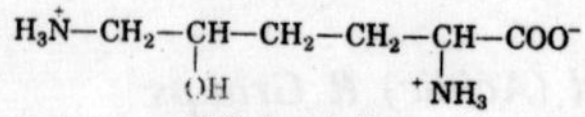

5-Hydroxylysine

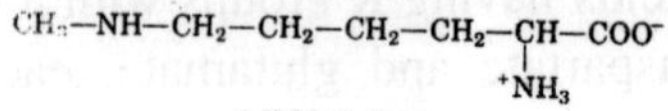

6-*N*-Methyllysine

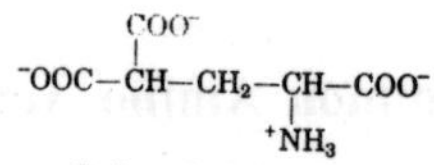

γ-Carboxyglutamate

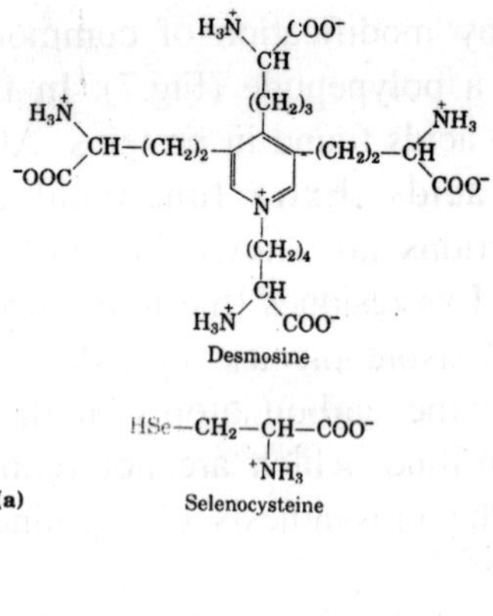

Desmosine

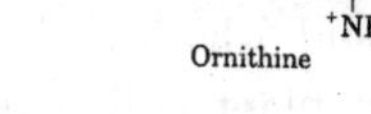

(a) Selenocysteine

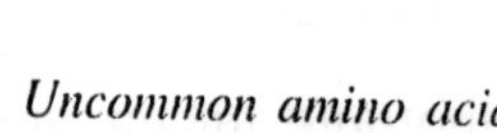

Ornithine

(b) Citrulline

Figure 7: Uncommon amino acids

four Lys residues, which is found in the fibrous protein elastin. Selenocysteine is a special case. This rare amino acid residue is introduced during protein synthesis rather than created through a postsynthetic modification. It contains selenium rather than the sulfur of cysteine. Actually derived from serine, selenocysteine is a constituent of just a few known proteins.

Some 300 additional amino acids have been found in cells. They have a variety of functions but are not constituents of proteins.

Ornithine and citrulline Amino Acids

When an amino acid is dissolved in water, it exists in solution as the dipolar ion, or zwitterion (German for "hybrid ion"), shown in Figure 8.

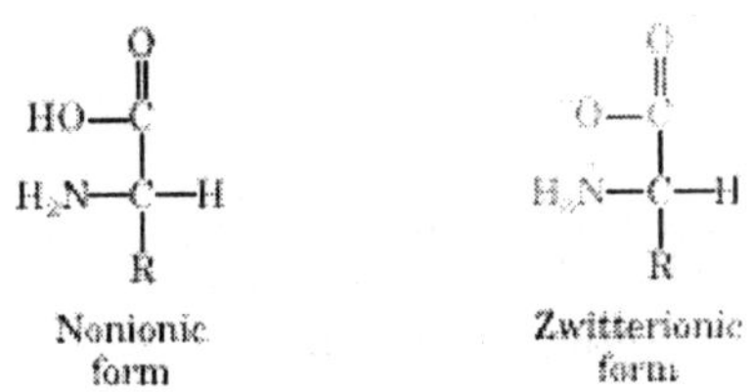

Figure 8: Nonionic and zwitterionic forms of amino acids.

A zwitterion can act as either an acid (proton donor):

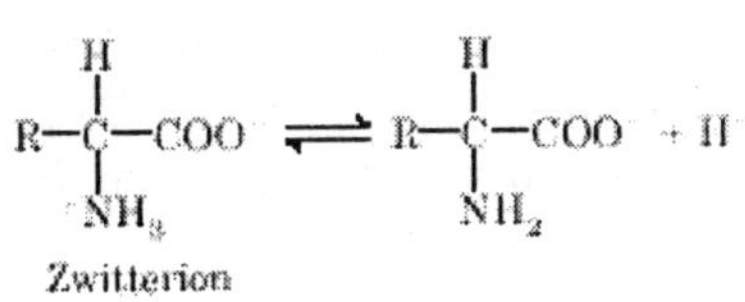

or a base (proton acceptor):

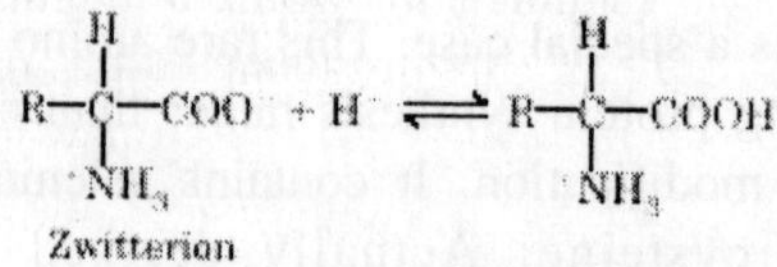

Substances having this dual nature are amphoteric and are often called ampholytes (from "amphoteric electrolytes"). A simple monoamino monocarboxylic α-amino acid, such as alanine, is a diprotic acid when fully protonated-it has two groups, the –COOH group and the $-NH_3^+$ group, that can yield protons:

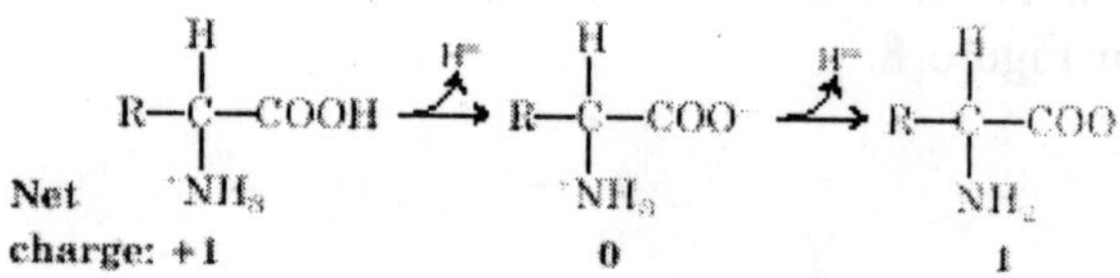

Absorption of Light by Molecules: The Lambert-Beer Law

A wide range of biomolecules absorb light at characteristic wavelengths, just as tryptophan absorbs light at 280nm. Measurement of light absorption by a spectrophotometer is used to detect and identify molecules and to measure their concentration in solution. The fraction of the incident light absorbed by a solution at a given wavelength is related to the thickness of the absorbing layer (path length) and the concentration of the absorbing species (Fig. 9). These two relationships are combined into the Lambert-Beer law,

$$\log \frac{I_0}{I} = \varepsilon c l$$

where I_0 is the intensity of the incident light, I is the intensity of the transmitted light, ε is the molar extinction coefficient (in units of liters per mole-centimeter), c is the concentration of the absorbing species (in moles per liter), and l is the path length of the lightabsorbing sample (in centimeters). The Lambert-Beer law assumes that the incident light is parallel and monochromatic (of a single wavelength) and that the solvent and solute molecules are randomly oriented.

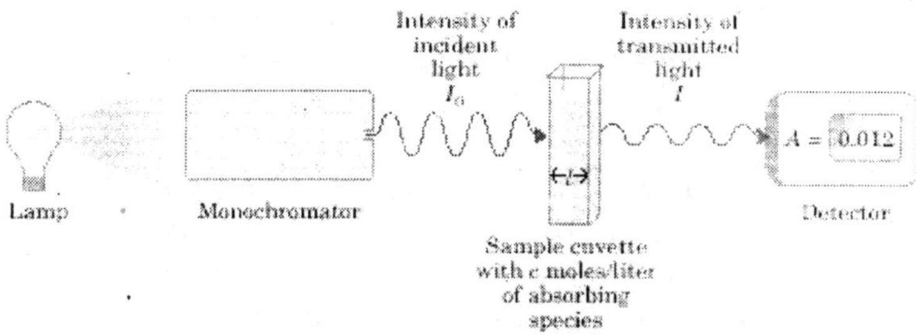

Figure 9: The principal components of a spectrophotometer

The expression log (I_0/I) is called the absorbance, designated A. It is important to note that each successive millimeter of path length of absorbing solution in a 1.0 cm cell absorbs not a constant amount but a constant fraction of the light that is incident upon it. However, with an absorbing layer of fixed path length, the absorbance, A, is directly proportional to the concentration of the absorbing solute.

The molar extinction coefficient varies with the nature of the absorbing compound, the solvent, and the wavelength, and also with pH if the light-absorbing species is in equilibrium with an ionization state that has different absorbance properties.

Amino Acids Have Characteristic Titration Curves

Acid-base titration involves the gradual addition or removal of protons. Figure 10 shows the titration curve of the diprotic form of glycine. The plot has two distinct stages, corresponding to deprotonation of two different groups on glycine. Each of the two stages resembles in shape the titration curve of a monoprotic acid, such as acetic acid, and can be analyzed in the same way. At very low pH, the predominant ionic species of glycine is the fully protonated form, $^{+}H_3N—CH_2—COOH$.

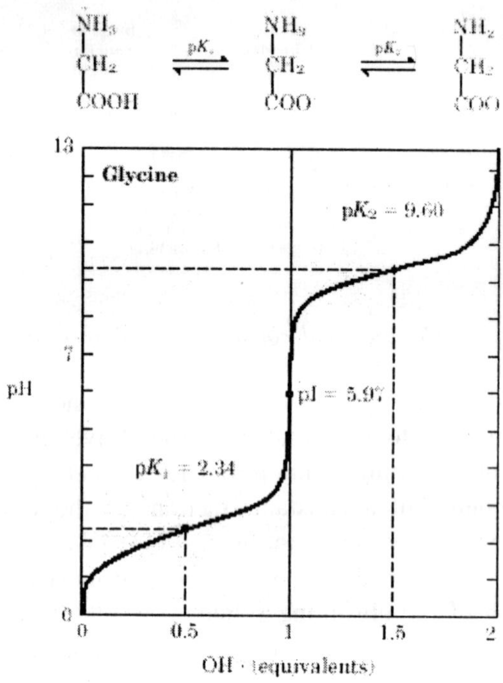

Figure 10: Titration of an amino acid.

At the midpoint in the first stage of the titration, in which the –COOH group of glycine loses its proton, equimolar concentrations of the proton-donor ($^{+}H_3N—CH_2—COOH$) and proton-acceptor (^+H_3N—CH_2—COO^-) species are present. At the midpoint of any titration, a point of inflection is reached where the pH is equal to the pK_a of the protonated group being titrated.

For glycine, the pH at the midpoint is 2.34, thus its —COOH group has a pK_a (labeled pK_1 in Fig 10) of 2.34.

As the titration proceeds, another important point is reached at pH 5.97. Here there is another point of inflection, at which removal of the first proton is essentially complete and removal of the second has just begun. At this pH glycine is present largely as the dipolar ion H_3N—CH_2—COO^-. We shall return to the significance of this inflection point in the titration curve (labeled pI in Fig. 10) shortly.

The second stage of the titration corresponds to the removal of a proton from the $-NH_3^+$ group of glycine. The pH at the midpoint of this stage is 9.60, equal to the pK_a (labeled pK_2 in Fig. 10) for the $-NH_3^+$ group.

The titration is essentially complete at a pH of about 12, at which point the predominant form of glycine is H_2N—CH_2—COO^-. From the titration curve of glycine we can derive several important pieces of information. First, it gives a quantitative measure of the pKa of each of the two ionizing groups: 2.34 for the —COOH group and 9.60 for the $-NH_3^+$ group. The carboxyl group of glycine is over 100 times more acidic (more easily ionized) than the carboxyl group of acetic acid, has a pK_a of 4.76-about average for a carboxyl group attached to an otherwise unsubstituted aliphatic hydrocarbon. The perturbed pK_a of glycine is caused by repulsion between the departing proton and the nearby positively charged amino group on the α-carbon atom, as described in Figure 11. The opposite charges on the resulting zwitterion are stabilizing, nudging the equilibrium farther to the right.

Similarly, the pK_a of the amino group in glycine is perturbed downward relative to the average pK_a of an amino group. This

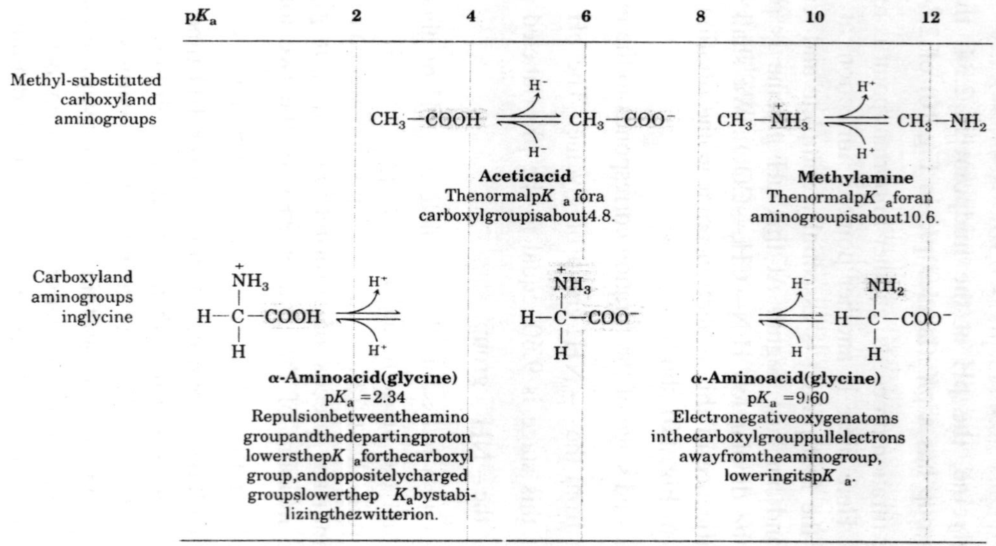

Figure 11: Effect of the chemical environment on pK_a.

effect is due partly to the electronegative oxygen atoms in the carboxyl groups, which tend to pull electrons toward them, increasing the tendency of the amino group to give up a proton. Hence, the α-amino group has a pK_a that is lower than that of an aliphatic amine such as methylamine (Fig. 11). In short, the pK_a of any functional group is greatly affected by its chemical environment, a phenomenon sometimes exploited in the active sites of enzymes to promote exquisitely adapted reaction mechanisms that depend on the perturbed pK_a values of proton donor/acceptor groups of specific residues.

The second piece of information provided by the titration curve of glycine is that this amino acid has two regions of buffering power. One of these is the relatively flat portion of the curve, extending for approximately 1 pH unit on either side of the first pK_a of 2.34, indicating that glycine is a good buffer near this pH. The other buffering zone is centered around pH 9.60. Within the buffering ranges of glycine, the Henderson-Hasselbalch equation can be used to calculate the proportions of proton-donor and proton-acceptor species of glycine required to make a buffer at a given pH.

Titration Curves Predict the Electric Charge of Amino Acids

Another important piece of information derived from the titration curve of an amino acid is the relationship between its net electric charge and the pH of the solution. At pH 5.97, the point of inflection between the two stages in its titration curve, glycine is present predominantly as its dipolar form, fully ionized but with no net electric charge (Fig. 10).

The characteristic pH at which the net electric charge is zero is called the isoelectric point or isoelectric pH, designated pI. For glycine, which has no ionizable group in its side chain, the isoelectric point is simply the arithmetic mean of the two pK_a values:

$$pI = \frac{1}{2}(pK_1 + pK_2) = \frac{1}{2}(2.34 + 9.60) = 5.97$$

As is evident in Figure 10, glycine has a net negative charge at any pH above its pI and will thus move toward the positive electrode (the anode) when placed in an electric field. At any pH below its pI, glycine has a net positive charge and will move toward the negative electrode (the cathode). The farther the pH of a glycine solution is from its isoelectric point, the greater the net electric charge of the population of glycine molecules. At pH 1.0, for example, glycine exists almost entirely as the form $^+H_3N—CH_2—COOH$, with a net positive charge of 1.0. At pH 2.34, where there is an equal mixture of $^+H_3N—CH_2—COOH$ and $^+H_3N—CH_2—CO^-$, the average or net positive charge is 0.5. The sign and the magnitude of the net charge of any amino acid at any pH can be predicted in the same way.

Amino Acids Differ in Their Acid-Base Properties

The shared properties of many amino acids permit some simplifying generalizations about their acid-base behaviors. First, all amino acids with a single α-amino group, a single α-carboxyl group, and an R group that does not ionize have titration curves resembling that of glycine (Fig. 10). These amino acids have very similar, although not identical, pK_a values: pK_a of the —COOH group in the range of 1.8 to 2.4, and pK_a of the $-NH_3^+$ group in the range of 8.8 to 11.0 (Table 1).

Second, amino acids with an ionizable R group have more complex titration curves, with three stages corresponding to the three possible ionization steps; thus they have three pKa values. The additional stage for the titration of the ionizable R group merges to some extent with the other two.

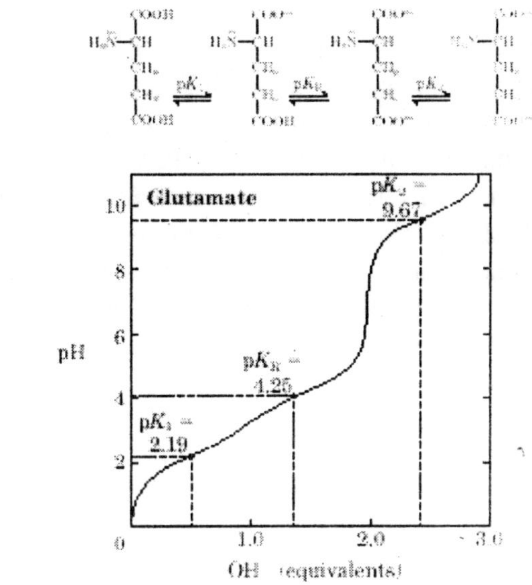

(a)

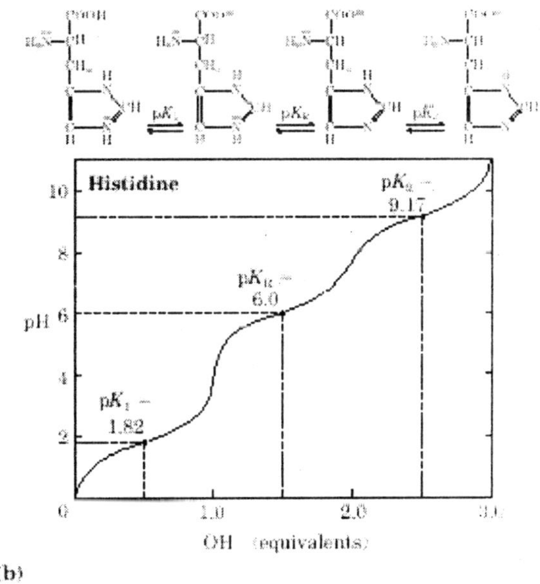

(b)

Figure 12: Titration curves for (a) glutamate and (b) histidine. The pK_a of the R group is designated here as pK_R.

The titration curves for two amino acids of this type, glutamate and histidine, are shown in Figure 12. The isoelectric points reflect the nature of the ionizing R groups present. For example, glutamate has a pI of 3.22, considerably lower than that of glycine. This is due to the presence of two carboxyl groups, which, at the average of their pK_a values (3.22), contribute a net charge of –1 that balances the –1 contributed by the amino group. Similarly, the pI of histidine, with two groups that are positively charged when protonated, is 7.59, much higher than that of glycine.

4

Peptides and Proteins

We now turn to polymers of amino acids, the peptides and proteins. Biologically occurring polypeptides range in size from small to very large, consisting of two or three to thousands of linked amino acid residues.

Two amino acid molecules can be covalently joined through a substituted amide linkage, termed a peptide bond, to yield a dipeptide. Such a linkage is formed by removal of the elements of water (dehydration) from the α-carboxyl group of one amino acid and the α-amino group of another (Fig.1).

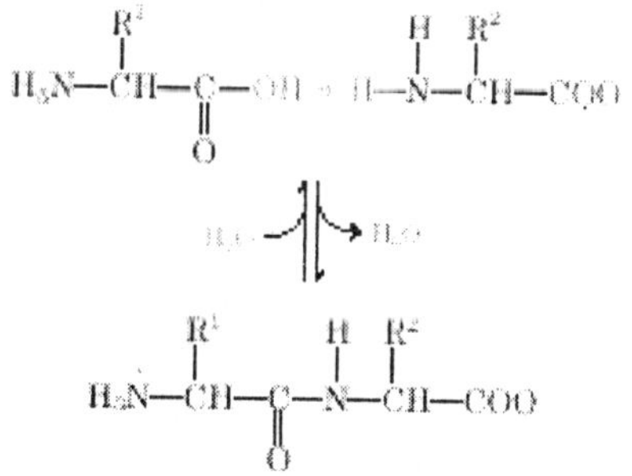

Figure 1. Formation of a peptide bond by condensation.

Peptide bond formation is an example of a condensation reaction, a common class of reactions in living cells. Under standard biochemical conditions, the equilibrium for the reaction shown in Figure 1 favors the amino acids over the dipeptide.

To make the reaction thermodynamically more favorable, the carboxyl group must be chemically modified or activated so that the hydroxyl group can be more readily eliminated. Three amino acids can be joined by two peptide bonds to form a tripeptide; similarly, amino acids can be linked to form tetrapeptides, pentapeptides, and so forth. When a few amino acids are joined in this fashion, the structure is called an oligopeptide.

When many amino acids are joined, the product is called a polypeptide. Proteins may have thousands of amino acid residues. Although the terms "protein" and "polypeptide" are sometimes used interchangeably, molecules referred to as polypeptides generally have molecular weights below 10,000, and those called proteins have higher molecular weights.

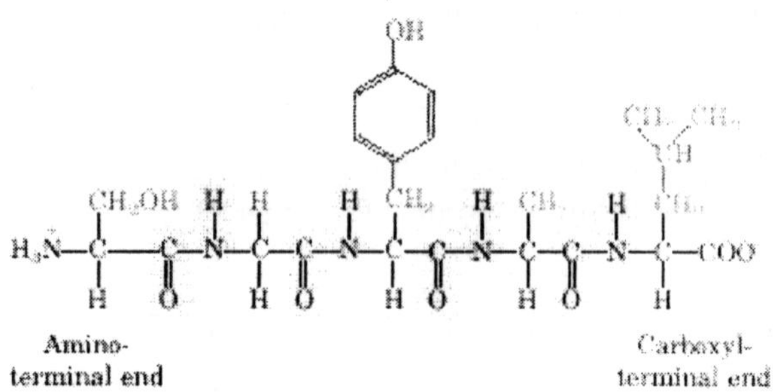

Figure 2. The pentapeptide serylglycyltyrosylalanylleucine, or Ser–Gly–Tyr–Ala–Leu.

Figure 2 shows the structure of a pentapeptide. As already noted, an amino acid unit in a peptide is often called a residue (the part left over after losing a hydrogen atom from its amino group and the hydroxyl moiety from its carboxyl group). In a peptide, the amino acid residue at the end with a free α-amino group is the amino-terminal (or *N*-terminal) residue; the residue at the other end, which has a free carboxyl group, is the carboxyl-terminal (*C*-terminal) residue. Although hydrolysis of a peptide bond is an exergonic reaction, it occurs slowly because of its high activation energy. As a result, the peptide bonds in proteins are quite stable, with an average half-life ($t_{1/2}$) of about 7 years under most intracellular conditions.

Ionization Behavior of Peptides

Peptides contain only one free α-amino group and one free α-carboxyl group, at opposite ends of the chain (Fig. 3). These groups ionize as they do in free amino acids, although the ionization constants are different because an oppositely charged group is no longer linked to the α carbon.

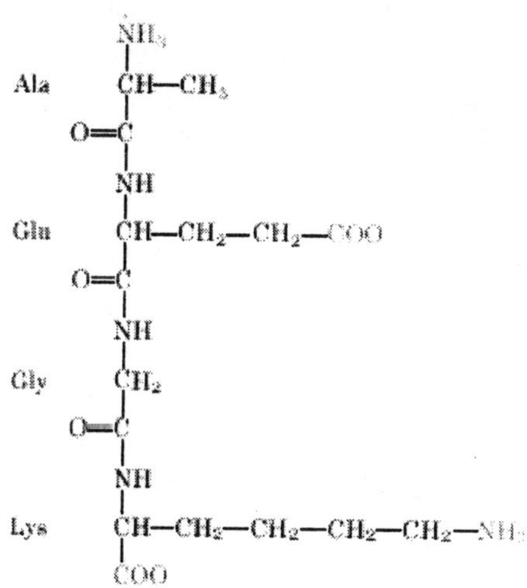

Figure 3. Alanylglutamylglycyllysine

The α-amino and α-carboxyl groups of all nonterminal amino acids are covalently joined in the peptide bonds, which do not ionize and thus do not contribute to the total acid-base behavior of peptides. However, However, the R groups of some amino acids can ionize, and in a peptide these contribute to the overall acid-base properties of the molecule (Fig. 3). Thus the acid-base behavior of a peptide can be predicted from its free α-amino and α-carboxyl groups as well as the nature and number of its ionizable R groups.

Like free amino acids, peptides have characteristic titration curves and a characteristic isoelectric pH (pI) at which they do not move in an electric field. These properties are exploited in some of the techniques used to separate peptides and proteins. It should be emphasized that the pKa value for an ionizable R group can change somewhat when an amino acid becomes a residue in a peptide. The loss of charge in the α-carboxyl and α-amino groups, the interactions with other peptide R groups, and other environmental factors can affect the pK_a.

The pK_a values for R groups listed in Table 1 can be a useful guide to the pH range in which a given group will ionize, but they cannot be strictly applied to peptides.

Active Peptides and Polypeptides

No generalizations can be made about the molecular weights of biologically active peptides and proteins in relation to their functions. Naturally occurring peptides range in length from two to many thousands of amino acid residues. Even the smallest peptides can have biologically important effects. Consider the commercially synthesized dipeptide *L*-aspartyl-*L*-phenylalanine methyl ester, the artificial sweetener better known as aspartame or NutraSweet.

Many small peptides exert their effects at very low concentrations. For example, a number of vertebrate hormones are small peptides. These include oxytocin (nine amino acid residues), which is secreted by the posterior pituitary and stimulates uterine contractions; bradykinin (nine residues), which inhibits inflammation of tissues; and thyrotropin-releasing factor (three residues), which is formed in the hypothalamus and stimulates the release of another hormone, thyrotropin, from the anterior pituitary gland.

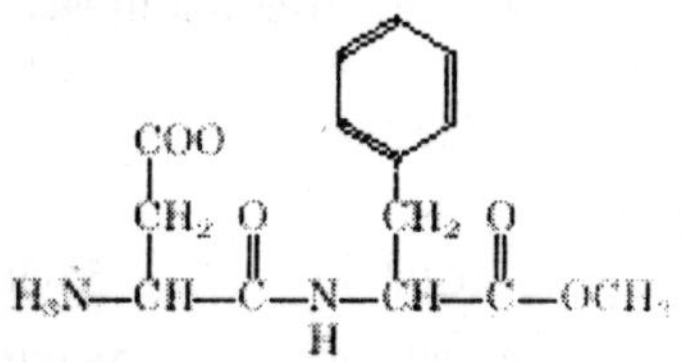

L-Aspartyl-L-phenylalanine methyl ester (aspartame)

Some extremely toxic mushroom poisons, such as amanitin, are also small peptides, as are many antibiotics. Slightly larger are small polypeptides and oligopeptides such as the pancreatic hormone insulin, which contains two polypeptide chains, one having 30 amino acid residues and the other 21. Glucagon, another pancreatic hormone, has 29 residues; it opposes the action of insulin.

Table 1. Molecular Data on Some Proteins

	Molecular weight	Number of residues	Number of polypeptide chains
Cytochrome c (human)	13,000	104	1
Ribonuclease A (bovine pancreas)	13,700	124	1
Lysozyme (chicken egg white)	13,930	129	1
Myoglobin (equine heart)	16,890	153	1
Chymotrypsin (bovine pancreas)	21,600	241	3

Chymotrypsinogen (bovine)	22,000	245	1
Hemoglobin (human)	64,500	574	4
Serum albumin (human)	68,500	609	1
Hexokinase (yeast)	102,000	972	2
RNA polymerase (E. coli)	450,000	4,158	5
Apolipoprotein B (human)	513,000	4,536	1
Glutamine synthetase (E. coli)	619,000	5,628	12
Titin (human)	2,993,000	26,926	1

Corticotropin is a 39-residue hormone of the anterior pituitary gland that stimulates the adrenal cortex. As Table 2 shows, lengths vary considerably. Human cytochrome c has 104 amino acid residues linked in a single chain; bovine chymotrypsinogen has 245 residues. At the extreme is titin, a constituent of vertebrate muscle, which has nearly 27,000 amino acid residues and a molecular weight of about 3,000,000. The vast majority of naturally occurring proteins are much smaller than this, containing fewer than 2,000 amino acid residues. Some proteins consist of a single polypeptide chain, but others, called multisubunit proteins, have two or more polypeptides associated noncovalently.

The individual polypeptide chains in a multisubunit protein may be identical or different. If at least two are identical the protein is said to be oligomeric, and the identical units (consisting of one or more polypeptide chains) are referred to as protomers. Hemoglobin, for example, has four polypeptide subunits: two identical α chains and two identical β chains, all four held together by noncovalent interactions. Each α subunit is paired in an identical way with a β subunit within the structure of this multisubunit protein, so that hemoglobin can be considered either a tetramer of four polypeptide subunits or a dimer of αβ protomers.

A few proteins contain two or more polypeptide chains linked covalently. For example, the two polypeptide chains of insulin are linked by disulfide bonds. In such cases, the individual polypeptides are not considered subunits but are commonly referred to simply as chains. We can calculate the approximate number of amino acid residues in a simple protein containing no other chemical constituents by dividing its molecular weight by 110.

Although the average molecular weight of the 20 common amino acids is about 138, the smaller amino acids predominate in most proteins. If we take into account the proportions in which the various amino acids occur in proteins, the average molecular weight of protein amino acids is nearer to 128. Because a molecule of water (M_r 18) is removed to create each peptide bond, the average molecular weight of an amino acid residue in a protein is about 128 – 18 = 110.

Amino Acid Compositions of Polypeptides

Hydrolysis of peptides or proteins with acid yields a mixture of free α-amino acids. When completely hydrolyzed, each type of protein yields a characteristic proportion or mixture of the different amino acids. The 20 common amino acids almost never occur in equal amounts in a protein. Some amino acids may occur only once or not at all in a given type of protein; others may occur in large numbers. Table 2 shows the composition of the amino acid mixtures obtained on complete hydrolysis of bovine cytochrome *c* and chymotrypsinogen, the inactive precursor of the digestive enzyme chymotrypsin. These two proteins, with very different functions, also differ significantly in the relative numbers of each kind of amino acid they contain.

Complete hydrolysis alone is not sufficient for an exact analysis of amino acid composition, however, because some side reactions occur during the procedure. For example, the amide bonds in the side chains of asparagine and glutamine are cleaved by acid treatment, yielding aspartate and glutamate, respectively. The side chain of tryptophan is almost completely degraded by acid hydrolysis, and small amounts of serine, threonine, and tyrosine are also lost. When a precise amino acid composition is required, biochemists use additional procedures to resolve the ambiguities that remain from acid hydrolysis.

Table 2. Amino Acid Composition of Two Proteins

	Number of residues per molecule of protein*	
Amino acid	Bovine cytochrome c	Bovine chymotrypsinogen
Ala	6	22
Arg	2	4
Asn	5	15
Asp	3	8
Cys	2	10
Gln	3	10
Glu	9	5
Gly	14	23
His	3	2
Ile	6	10
Leu	6	19
Lys	18	14
Met	2	2
Phe	4	6
Pro	4	9
Ser	1	28
Thr	8	23

Trp	1	8
Tyr	4	4
Val	3	23
Total	104	245

*In some common analyses, such as acid hydrolysis, Asp and Asn are not readily distin-guished from each other and are together designated Asx (or B). Similarly, when Glu and Gln cannot be distinguished, they are together designated Glx (or Z). In addition, Trp is destroyed. Additional procedures must be employed to obtain an accurate assessment of complete amino acid content.

Some Proteins Containing Chemical Groups Other Than Amino Acids

Many proteins, for example the enzymes ribonuclease A and chymotrypsinogen, contain only amino acid residues and no other chemical constituents; these are considered simple proteins. However, some proteins contain permanently associated chemical components in addition to amino acids; these are called conjugated proteins. The non-amino acid part of a conjugated protein is usually called its prosthetic group.

Table 3. Conjugated Proteins

Class	Prosthetic group	Example
Lipoproteins	Lipids	β_1-Lipoprotein of blood
Glycoproteins	Carbohydrates	Immuno globulin G
Phosphoproteins	Phosphate groups	Casein of milk
Hemoproteins	Heme (iron porphyrin)	Hemo globin
Flavoproteins	Flavin nucleotides	Succinate dehydro genase

Metalloproteins	Iron	Ferritin
	Zinc	Alcohol dehydro genase
	Calcium	Calmo dulin
	Molybdenum	Dinitro genase
	Copper	Plasto cyanin

Conjugated proteins are classified on the basis of the chemical nature of their prosthetic groups (Table 4); for example, lipoproteins contain lipids, glycoproteins contain sugar groups, and metalloproteins contain a specific metal. A number of proteins contain more than one prosthetic group. Usually the prosthetic group plays an important role in the protein's biological function.

Levels of Protein Structure

For large macromolecules such as proteins, the tasks of describing and understanding structure are approached at several levels of complexity, arranged in a kind of conceptual hierarchy. Four levels of protein structure are commonly defined (Fig. 4). A description of all covalent bonds (mainly peptide bonds and disulfide bonds) linking amino acid residues in a polypeptide chain is its primary structure. The most important element of primary structure is the sequence of amino acid residues.

Secondary structure refers to particularly stable arrangements of amino acid residues giving rise to recurring structural patterns. Tertiary structure describes all aspects of the three-dimensional folding of a polypeptide. When a protein has two or more polypeptide subunits, their arrangement in space is referred to as quaternary structure.

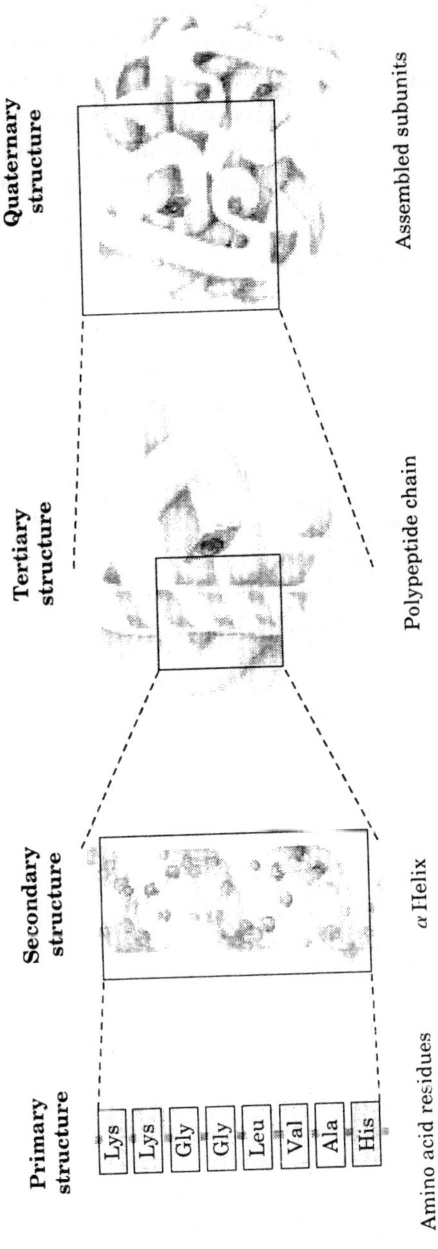

Figure 4. Levels of structure in proteins.

Separated and Purified Proteins

A pure preparation is essential before a protein's properties and activities can be determined. Given that cells contain thousands of different kinds of proteins, how can one protein be purified? Methods for separating proteins take advantage of properties that vary from one protein to the next, including size, charge, and binding properties.

The source of a protein is generally tissue or microbial cells. The first step in any protein purification procedure is to break open these cells, releasing their proteins into a solution called a crude extract. If necessary, differential centrifugation can be used to prepare subcellular fractions or to isolate specific organelles.

Once the extract or organelle preparation is ready, various methods are available for purifying one or more of the proteins it contains. Commonly, the extract is subjected to treatments that separate the proteins into different fractions based on a property such as size or charge, a process referred to as fractionation. Early fractionation steps in a purification utilize differences in protein solubility, which is a complex function of pH, temperature, salt concentration, and other factors. The solubility of proteins is generally lowered at high salt concentrations, an effect called "salting out." The addition of a salt in the right amount can selectively precipitate some proteins, while others remain in solution. Ammonium sulfate ($(NH_4)_2SO_4$) is often used for this purpose because of its high solubility in water. A solution containing the protein of interest often must be further altered before subsequent purification steps are possible. For example, dialysis is a procedure that separates proteins from solvents by taking advantage of the proteins' larger size.

The partially purified extract is placed in a bag or tube made of a semipermeable membrane. When this is suspended in a much larger volume of buffered solution of appropriate ionic strength, the membrane allows the exchange of salt and buffer but not proteins. Thus dialysis retains large proteins within the membranous bag or tube while allowing the concentration of other solutes in the protein preparation to change until they come into equilibrium with the solution outside the membrane. Dialysis might be used, for example, to remove ammonium sulfate from the protein preparation.

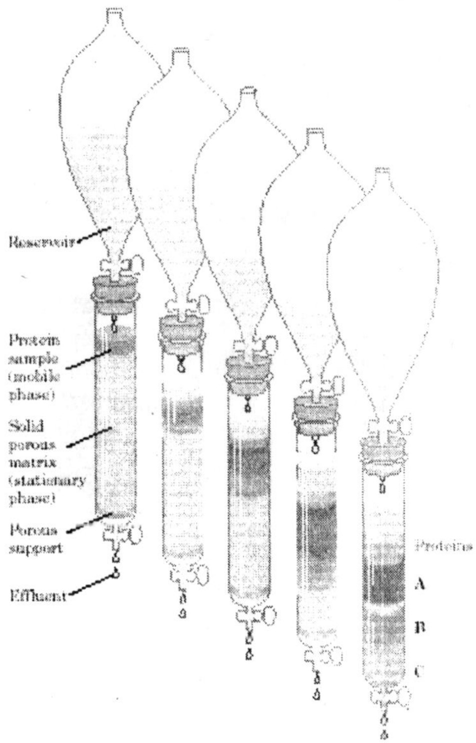

Figure 5. Column chromatography

The most powerful methods for fractionating proteins make use of column chromatography, which takes advantage of differences in protein charge, size, binding affinity, and other properties. A porous solid material with appropriate chemical properties (the stationary phase) is held in a column, and a buffered solution (the mobile phase) percolates through it. The protein-containing solution, layered on the top of the column, percolates through the solid matrix as an ever-expanding band within the larger mobile phase (Fig. 6).

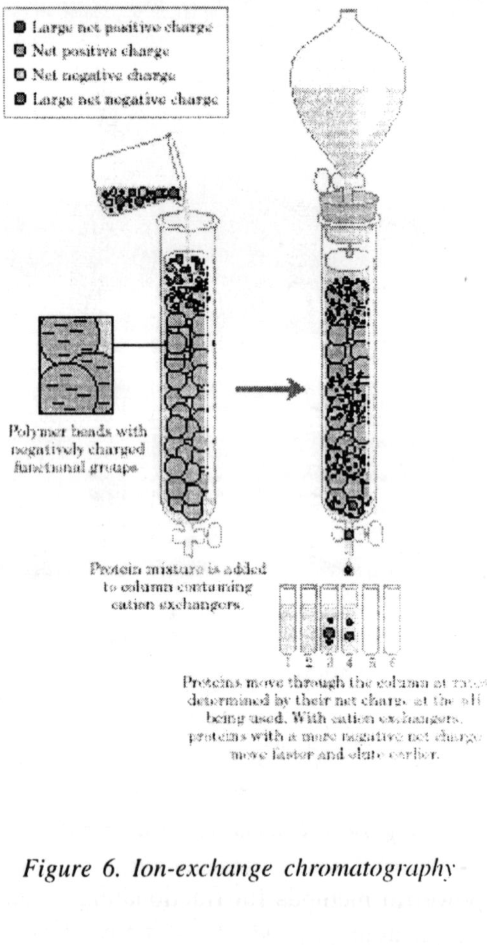

Figure 6. Ion-exchange chromatography

Individual proteins migrate faster or more slowly through the column depending on their properties. For example, in cation-exchange chromatography (Fig. 7), the solid matrix has negatively charged groups. In the mobile phase, proteins with a net positive charge migrate through the matrix more slowly than those with a net negative charge, because the migration of the former is retarded more by interaction with the stationary phase. The two types of protein can separate into two distinct bands. The expansion of the protein band in the mobile phase (the protein solution) is caused both by separation of proteins with different properties and by diffusional spreading.

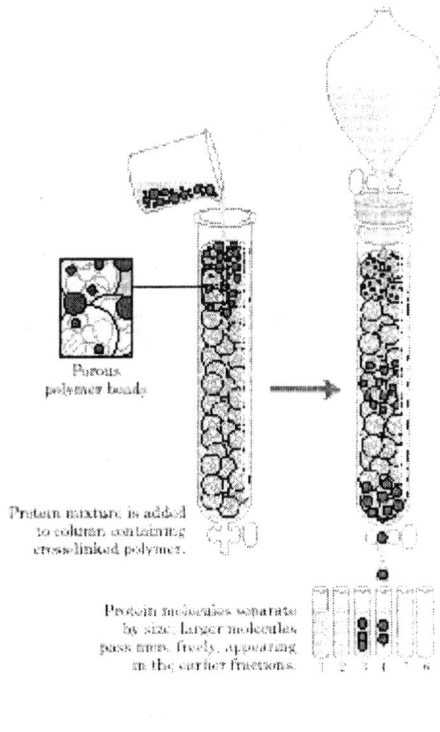

Figure 7. Size-exclusion chromatography

As the length of the column increases, the resolution of two types of protein with different net charges generally improves. However, the rate at which the protein solution can flow through the column usually decreases with column length. And as the length of time spent on the column increases, the resolution can decline as a result of diffusional spreading within each protein band.

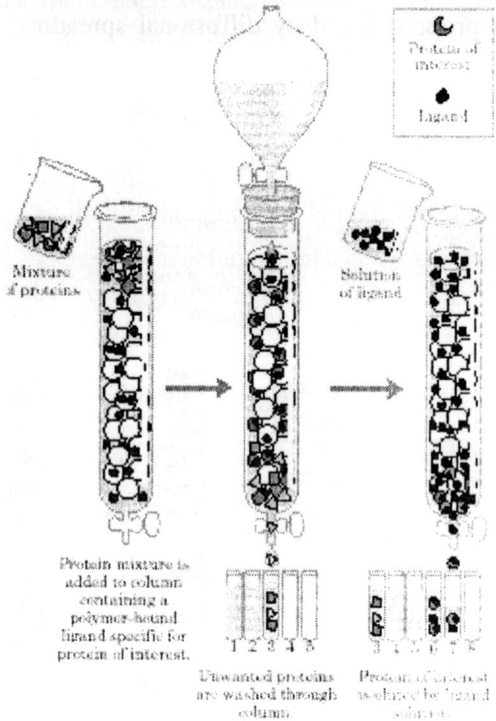

Figure 8. Affinity chromatography

Sizeexclusion chromatography separates proteins according to size. In this method, large proteins emerge from the column sooner than small ones-a somewhat counterintuitive result. The solid phase consists of beads with engineered pores or cavities of a particular size. Large proteins cannot enter the cavities, and so take a short (and rapid) path through the column, around the beads. Small proteins enter the cavities, and migrate through the column more slowly as a result.

Affinity chromatography is based on the binding affinity of a protein. The beads in the column have a covalently attached chemical group. A protein with affinity for this particular chemical group will bind to the beads in the column, and its migration will be retarded as a result.

A modern refinement in chromatographic methods is HPLC, or high-performance liquid chromatography. HPLC makes use of high-pressure pumps that speed the movement of the protein molecules down the column, as well as higher-quality chromatographic materials that can withstand the crushing force of the pressurized flow. By reducing the transit time on the column, HPLC can limit diffusional spreading of protein bands and thus greatly improve resolution. The approach to purification of a protein that has not previously been isolated is guided both by established precedents and by common sense. In most cases, several different methods must be used sequentially to purify a protein completely. The choice of method is somewhat empirical, and many protocols may be tried before the most effective one is found. Trial and error can often be minimized by basing the procedure on purification techniques developed for similar proteins.

Published purification protocols are available for many thousands of proteins. Common sense dictates that inexpensive procedures such as salting out be used first, when the total volume and the number of contaminants are greatest. Chromatographic methods are often impractical at early stages. because the amount of chromatographic medium needed increases with sample size. As each purification step is completed, the sample size generally becomes smaller, making it feasible to use more sophisticated (and expensive) chromatographic procedures at later stages.

Table 4. A Purification Table for a Hypothetical Enzyme

Procedure or step	Fraction volume (ml)	Total protein (mg)	Activity (units)	Specific activity (units/mg)
1. Crude cellular extract	1,400	10,000	100,000	10
2. Precipitation with ammonium sulfate	280	3,000	96,000	32
3. Ion-exchange chromatography	90	400	80,000	200
4. Size-exclusion chromatography	80	100	60,000	600
5. Affinity chromatography	6	3	45,000	15,000

Separated and Characterized Proteins by Electrophoresis

Another important technique for the separation of proteins is based on the migration of charged proteins in an electric field, a process called electrophoresis. These procedures are not generally used to purify proteins in large amounts, because simpler alternatives are usually available and electrophoretic methods often adversely affect the structure and thus the function of proteins.

Electrophoresis is, however, especially useful as an analytical method. Its advantage is that proteins can be visualized as well as separated, permitting a researcher to estimate quickly the number of different proteins in a mixture or the degree of purity of a particular protein preparation. Also, electrophoresis allows determination of crucial properties of a protein such as its isoelectric point and approximate molecular weight.

Electrophoresis of proteins is generally carried out in gels made up of the cross-linked polymer polyacrylamide. The polyacrylamide gel acts as a molecular sieve, slowing the migration of proteins approximately in proportion to their charge-to-mass ratio. Migration may also be affected by protein shape. In electrophoresis, the force moving the macromolecule is the electrical potential, *E*. The electrophoretic mobility of the molecule, μ, is the ratio of the velocity of the particle particle molecule, V, to the electrical potential. Electrophoretic mobility is also equal to the net charge of the molecule, *Z*, divided by the frictional coefficient, f, which reflects in part a protein's shape. Thus:

$$\mu = \frac{V}{E} = \frac{Z}{f}$$

The migration of a protein in a gel during electrophoresis is therefore a function of its size and its shape. An electrophoretic method commonly employed for estimation of purity and molecular weight makes use of the detergent sodium dodecyl sulfate (SDS).

Sodium dodecyl sulfate (SDS)

SDS binds to most proteins in amounts roughly proportional to the molecular weight of the protein, about one molecule of SDS for every two amino acid residues. The bound SDS contributes a large net negative charge, rendering the intrinsic charge of the protein insignificant and conferring on each protein a similar charge-to-mass ratio.

In addition, the native conformation of a protein is altered when SDS is bound, and most proteins assume a similar shape.

Electrophoresis in the presence of SDS therefore separates proteins almost exclusively on the basis of mass (molecular weight), with smaller polypeptides migrating more rapidly.

(a)

(b)

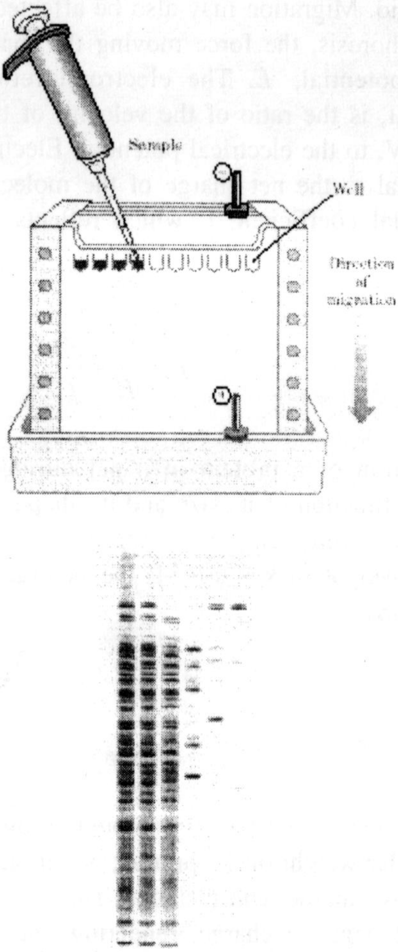

Figure 9. Electrophoresis. (a) Different samples are loaded in wells or depressions at the top of the polyacrylamide gel. (b) Proteins can be visualized after electrophoresis by treating the gel with a stain such as Coomassie blue, which binds to the proteins but not to the gel itself.

After electrophoresis, the proteins are visualized by adding a dye such as Coomassie blue, which binds to proteins but not to the gel itself (Fig. 9b). Thus, a researcher can monitor the progress of a protein purification procedure as the number of protein bands visible on the gel decreases after each new fractionation step.

When compared with the positions to which proteins of known molecular weight migrate in the gel, the position of an unidentified protein can provide an excellent measure of its molecular weight (Fig. 10). If the protein has two or more different subunits, the subunits will generally be separated by the SDS treatment and a separate band will appear for each.

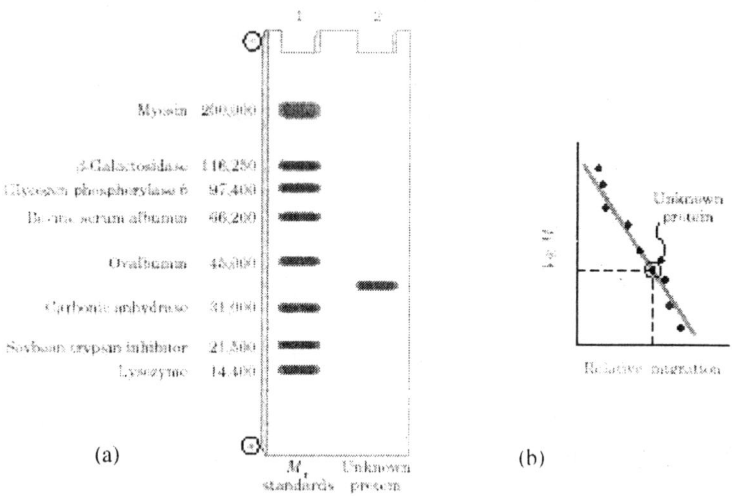

Figure 10: Estimating the molecular weight of a protein. (a) Standard proteins of known molecular weight are subjected to electrophoresis (lane 1). (b) A plot of log M_r of the marker proteins versus relative migration during electrophoresis is linear, which allows the molecular weight of the unknown protein to be read from the graph.

Isoelectric focusing is a procedure used to determine the isoelectric point (pI) of a protein (Fig. 11). A pH gradient is established by allowing a mixture of low molecular weight organic acids and bases to distribute themselves in an electric field generated across the gel. When a protein mixture is applied, each protein migrates until it reaches the pH that matches its pI (Table 5). Proteins with different isoelectric points are thus distributed differently throughout the gel.

Table 5. The Isoelectric Points of Some Proteins

Protein	*pI*
Pepsin	<1.0
Egg albumin	4.6
Serum albumin	4.9
Urease	5.0
β Lactoglobulin	5.2
Hemoglobin	6.8
Myoglobin	7.0
Chymotrypsinogen	9.5
Cytochrome c	10.7
Lysozyme	11.0

Combining isoelectric focusing and SDS electrophoresis sequentially in a process called two-dimensional electrophoresis permits the resolution of complex mixtures of proteins (Fig. 11). This is a more sensitive analytical method than either electrophoretic method alone. Two-dimensional electrophoresis separates proteins of identical molecular weight that differ in pI, or proteins with similar pI values but different molecular weights.

Unseparated Proteins

To purify a protein, it is essential to have a way of detecting and quantifying that protein in the presence of many other proteins at each stage of the procedure. Often, purification must proceed in the absence of any information about the size and physical properties of the protein or about the fraction of the total protein mass it represents in the extract.

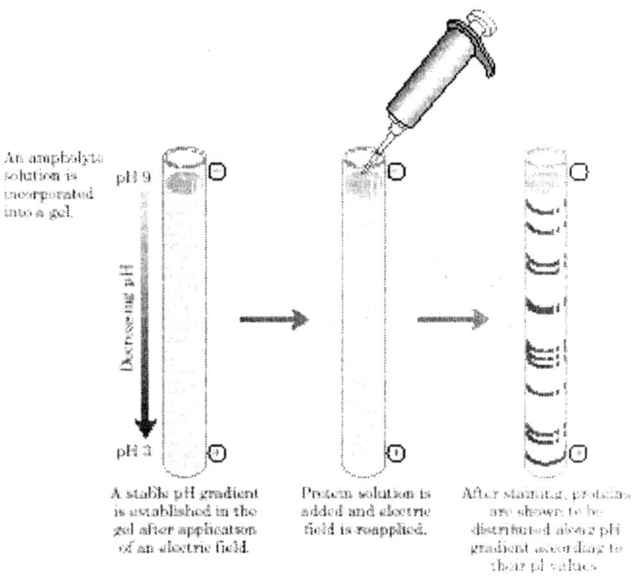

Figure 11. Isoelectric focusing

For proteins that are enzymes, the amount in a given solution or tissue extract can be measured, or assayed, in terms of the catalytic effect the enzyme produces-that is, the increase in the rate at which its substrate is converted to reaction products when the enzyme is present. For this purpose one must know

(1) the overall equation of the reaction catalyzed,

(2) an analytical procedure for determining the disappearance of the substrate or the appearance of a reaction product,

(3) whether the enzyme requires cofactors such as metal ions or coenzymes,

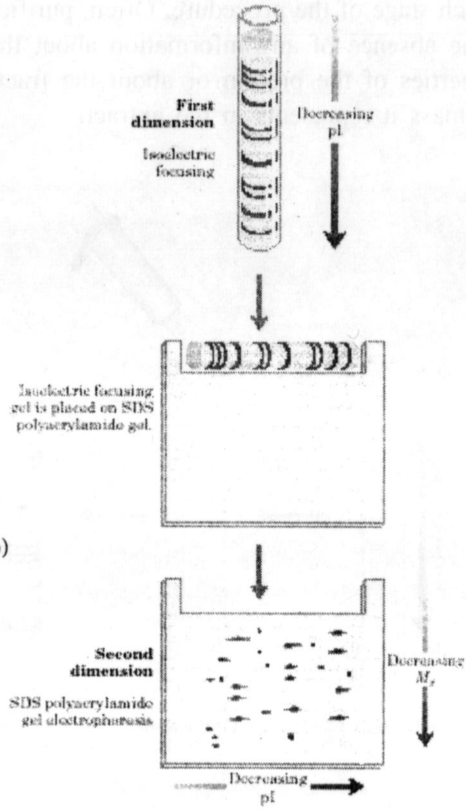

(b)

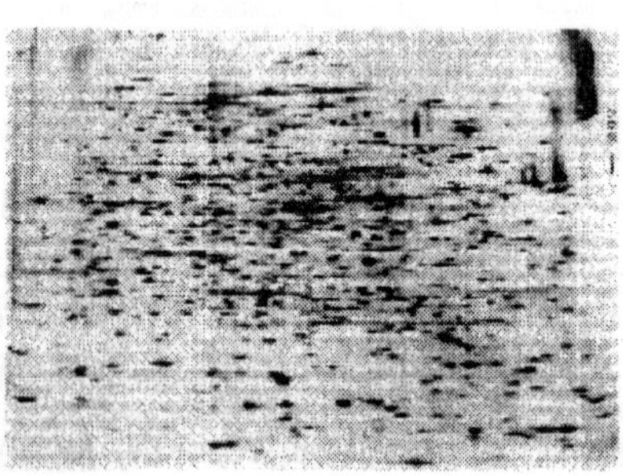

(a)

Figure 12. Two-dimensional electrophoresis. (a) Proteins are first separated by isoelectric focusing in a cylindrical gel. (b) More than 1,000 different proteins from E. coli can be resolved using this technique.

(4) the dependence of the enzyme activity on substrate concentration,

(5) the optimum pH, and

(6) a temperature zone in which the enzyme is stable and has high activity.

Enzymes are usually assayed at their optimum pH and at some convenient temperature within the range 25 to 38°C. Also, very high substrate concentrations are generally used so that the initial reaction rate, measured experimentally, is proportional to enzyme concentration.

By international agreement, 1.0 unit of enzyme activity is defined as the amount of enzyme causing transformation of 1.0 μmol of substrate per minute at 25°C under optimal conditions of measurement. The term activity refers to the total units of enzyme in a solution. The specific activity is the number of enzyme units per milligram of total protein (Fig. 13).

The specific activity is a measure of enzyme purity: it increases during purification of an enzyme and becomes maximal and constant when the enzyme is pure. After each purification step, the activity of the preparation (in units of enzyme activity) is assayed, the total amount of protein is determined independently, and the ratio of the two gives the specific activity. Activity and total protein generally decrease with each step.

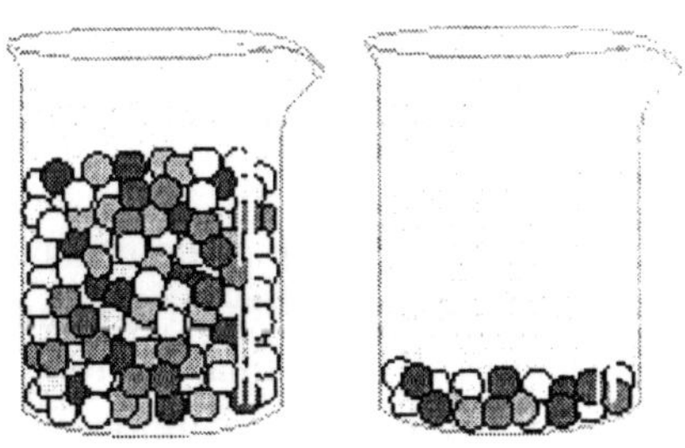

Figure 13. Activity versus specific activity

Activity decreases because some loss always occurs due to inactivation or nonideal interactions with chromatographic materials or other molecules in the solution. Total protein decreases because the objective is to remove as much unwanted or nonspecific protein as possible. In a successful step, the loss of nonspecific protein is much greater than the loss of activity; therefore, specific activity increases even as total activity falls. The data are then assembled in a purification table similar to Table 4. A protein is generally considered pure when further purification steps fail to increase specific activity and when only a single protein species can be detected (for example, by electrophoresis).

For proteins that are not enzymes, other quantification methods are required. Transport proteins can be assayed by their binding to the molecule they transport, and hormones and toxins by the biological effect they produce; for example, growth hormones will stimulate the growth of certain cultured cells. Some structural proteins represent such a large fraction of a tissue mass that they can be readily extracted and purified without a functional assay. The approaches are as varied as the proteins themselves.

Covalent Structure of Proteins

Purification of a protein is usually only a prelude to a detailed biochemical dissection of its structure and function. What is it that makes one protein an enzyme, another a hormone, another a structural protein, and still another an antibody? How do they differ chemically?

The differences in primary structure can be especially informative. Each protein has a distinctive number and sequence of amino acid residues. The primary structure of a protein determines how it folds up into a unique three-dimensional structure, and this in turn determines the function of the protein.

Importance of Amino Acid Sequence

The bacterium Escherichia coli produces more than 3,000 different proteins; a human produces 25,000 to 35,000. In both cases, each type of protein has a unique three-dimensional structure and this structure confers a unique function. Each type of protein also has a unique amino acid sequence. Intuition suggests that the amino acid sequence must play a fundamental role in determining the three-dimensional structure of the protein, and ultimately its function, but is this supposition correct? correct?

A quick survey of proteins and how they vary in amino acid sequence provides a number of empirical clues that help substantiate the important relationship between amino acid sequence and biological function. First, proteins with different functions always have different amino acid sequences. Second, thousands of human genetic diseases have been traced to the production of defective proteins. Perhaps one-third of these proteins are defective because of a single change in their amino acid sequence; hence, if the primary structure is altered, the function of the protein may also be changed.

Finally, on comparing functionally similar proteins from different species, these proteins often have similar amino acid sequences. An extreme case is ubiquitin, a 76-residue protein involved in regulating the degradation of other proteins. The amino acid sequence of ubiquitin is identical in species as disparate as fruit flies and humans.

An estimated 20% to 30% of the proteins in humans are polymorphic, having amino acid sequence variants in the human population. Many of these variations in sequence have little or no effect on the function of the protein. Furthermore, proteins that carry out a broadly similar function in distantly related species can differ greatly in overall size and amino acid sequence.

Although the amino acid sequence in some regions of the primary structure might vary considerably without affecting biological function, most proteins contain crucial regions that are essential to their function and whose sequence is therefore conserved. The fraction of the overall sequence that is critical varies from protein to protein, complicating the task of relating sequence to three-dimensional structure, and structure to function. Before we can consider this problem further, however, we must examine how sequence information is obtained.

Two major discoveries in 1953 were of crucial importance in the history of biochemistry. In that year James D. Watson and Francis Crick deduced the double-helical structure of DNA and proposed a structural basis for its precise replication. Their proposal illuminated the molecular reality behind the idea of a gene.

In that same year, Frederick Sanger worked out the sequence of amino acid residues in the polypeptide chains of the hormone insulin, surprising many researchers who had long thought that elucidation of the amino acid sequence of a polypeptide would be a hopelessly difficult task. It quickly became evident that the nucleotide sequence in DNA and the amino acid sequence in proteins were somehow related. Barely a decade after these discoveries, the role of the nucleotide sequence of DNA in determining the amino acid sequence of protein molecules was revealed.

An enormous number of protein sequences can now be derived indirectly from the DNA sequences in the rapidly growing genome databases. However, many are still deduced by traditional methods of polypeptide sequencing. The amino acid sequences of thousands of different proteins from many species have been determined using principles first developed by Sanger.Chemical protein sequencing now complements a growing list of newer methods, providing multiple avenues to obtain amino acid sequence data. Such data are now critical to every area of biochemical investigation.

Short Polypeptides Are Sequenced Using Automated Procedures

Various procedures are used to analyze protein primary structure. Several protocols are available to label and identify the amino-terminal amino acid residue (Fig. 14a). Sanger developed the reagent 1-fluoro-2,4-dinitrobenzene (FDNB) for this purpose; other reagents used to label the amino-terminal residue, dansyl chloride and dabsyl chloride, yield derivatives that are more easily detectable than the dinitrophenyl derivatives. After the amino-terminal residue is labeled with one of these reagents, the polypeptide is hydrolyzed to its constituent amino acids and the labeled amino acid is identified.

Because the hydrolysis stage destroys the polypeptide, this procedure cannot be used to sequence a polypeptide beyond its amino-terminal residue. However, it can help determine the number of chemically distinct polypeptides in a protein, provided each has a different amino-terminal residue. For example, two residues-Phe and Gly-would be labeled if insulin were subjected to this procedure.

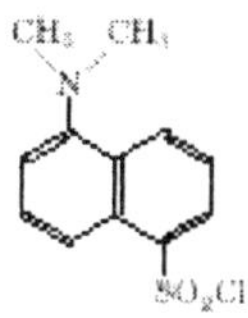

Dansyl chloride

Dabsyl chloride

To sequence an entire polypeptide, a chemical method devised by Pehr Edman is usually employed. The Edman degradation procedure labels and removes only the amino-terminal residue from a peptide, leaving all other peptide bonds intact (Fig. 14b). The peptide is reacted with phenylisothiocyanate under mildly

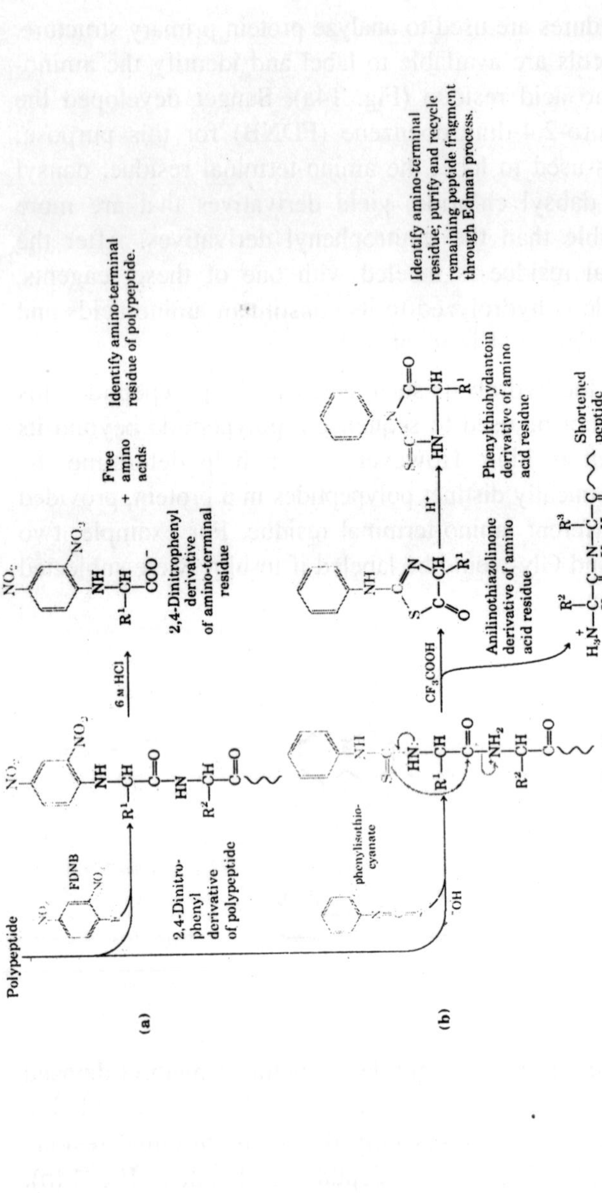

Figure 14. Steps in sequencing a polypeptide. (a) Identification of the amino-terminal residue can be the first step in sequencing a polypeptide. (b) The Edman degradation procedure reveals the entire sequence of a peptide.

alkaline conditions, which converts the aminoterminal amino acid to a phenylthiocarbamoyl (PTC) adduct.

The peptide bond next to the PTC adduct is then cleaved in a step carried out in anhydrous trifluoroacetic acid, with removal of the amino-terminal amino acid as an anilinothiazolinone derivative. The derivatized amino acid is extracted with organic solvents, converted to the more stable phenylthiohydantoin derivative by treatment with aqueous acid, and then identified.

The use of sequential reactions carried out under first basic and then acidic conditions provides control over the entire process. Each reaction with the aminoterminal amino acid can go essentially to completion without affecting any of the other peptide bonds in the peptide. After removal and identification of the aminoterminal residue, the new amino-terminal residue so exposed can be labeled, removed, and identified through the same series of reactions. This procedure is repeated until the entire sequence is determined. The Edman degradation is carried out on a machine, called a sequenator, that mixes reagents in the proper proportions, separates the products, identifies them, and records the results. These methods are extremely sensitive. Often, the complete amino acid sequence can be determined starting with only a few micrograms of protein.

The length of polypeptide that can be accurately sequenced by the Edman degradation depends on the efficiency of the individual chemical steps. The overall accuracy of amino acid sequencing generally declines as the length of the polypeptide increases. The very large polypeptides found in proteins must be broken down into smaller pieces to be sequenced efficiently. There are several steps in this process.

First, the protein is cleaved into a set of specific fragments by chemical or enzymatic methods. If any disulfide bonds are present, they must be broken. Each fragment is purified, then sequenced by the Edman procedure. Finally, the order in which the fragments appear in the original protein is determined and disulfide bonds (if any) are located.

Breaking Disulfide Bonds Disulfide bonds interfere with the sequencing procedure. A cystine residue that has one of its peptide bonds cleaved by the Edman procedure may remain attached to another polypeptide strand via its disulfide bond. Disulfide bonds also interfere with the enzymatic or chemical cleavage of the polypeptide. Two approaches to irreversible breakage of disulfide bonds are outlined in Figure 15.

Cleaving the Polypeptide Chain Several methods can be used for fragmenting the polypeptide chain. Enzymes called proteases catalyze the hydrolytic cleavage of peptide bonds. Some proteases cleave only the peptide bond adjacent to particular amino acid residues and thus fragment a polypeptide chain in a predictable and reproducible way. A number of chemical reagents also cleave the peptide bond adjacent to specific residues. Among proteases, the digestive enzyme trypsin catalyzes the hydrolysis of only those peptide bonds in which the carbonyl group is contributed by either a Lys or an Arg residue, regardless of the length or amino acid sequence of the chain. The number of smaller peptides produced by trypsin cleavage can thus be predicted from the total number of Lys or Arg residues in the original polypeptide, as determined by hydrolysis of an intact sample (Fig. 16).

A polypeptide with five Lys and/or Arg residues will usually yield six smaller peptides on cleavage with trypsin. Moreover, all except one of these will have a carboxyl-terminal Lys or Arg. The fragments produced by trypsin (or other enzyme or chemical) action are then separated by chromatographic or electrophoretic methods. Sequencing of Peptides Each peptide fragment resulting from the action of trypsin is sequenced separately by the Edman procedure.

Ordering Peptide Fragments

The order of the "trypsin fragments" in the original polypeptide chain must now be determined. Another sample of the intact polypeptide is cleaved into fragments using a different enzyme

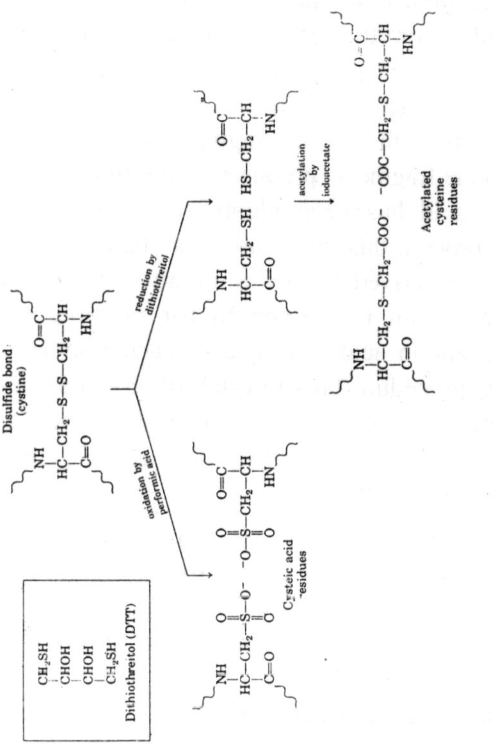

Figure 15. Breaking disulfide bonds in proteins.

or reagent, one that cleaves peptide bonds at points other than those cleaved by trypsin. For example, cyanogen bromide cleaves only those peptide bonds in which the carbonyl group is contributed by Met. The fragments resulting from this second procedure are then separated and sequenced as before.

The amino acid sequences of each fragment obtained by the two cleavage procedures are examined, with the objective of finding peptides from the second procedure whose sequences establish continuity, because of overlaps, between the fragments obtained by the first cleavage procedure (Fig. 16). Overlapping peptides obtained from the second fragmentation yield the correct order of the peptide fragments produced in the first. If the amino-terminal amino acid has been identified before the original cleavage of the protein, this information can be used to establish which fragment is derived from the amino terminus. The two sets of fragments can be compared for possible errors in determining the amino acid sequence of each fragment. If the second cleavage procedure fails to establish continuity between all peptides from the first cleavage, a third or even a fourth cleavage method must be used to obtain a set of peptides that can provide the necessary overlap(s).

Locating Disulfide Bonds If the primary structure includes disulfide bonds, their locations are determined in an additional step after sequencing is completed. A sample of the protein is again cleaved with a reagent such as trypsin, this time without first breaking the disulfide bonds. The resulting peptides are separated by electrophoresis and compared with the original set of peptides generated by trypsin. For each disulfide bond, two of the original peptides will be missing and a new, larger peptide will appear. The two missing peptides represent the regions of the intact polypeptide that are linked by the disulfide bond.

New methods for Amino Acid Sequences

The approach outlined above is not the only way to determine amino acid sequences. New methods based on mass spectrometry

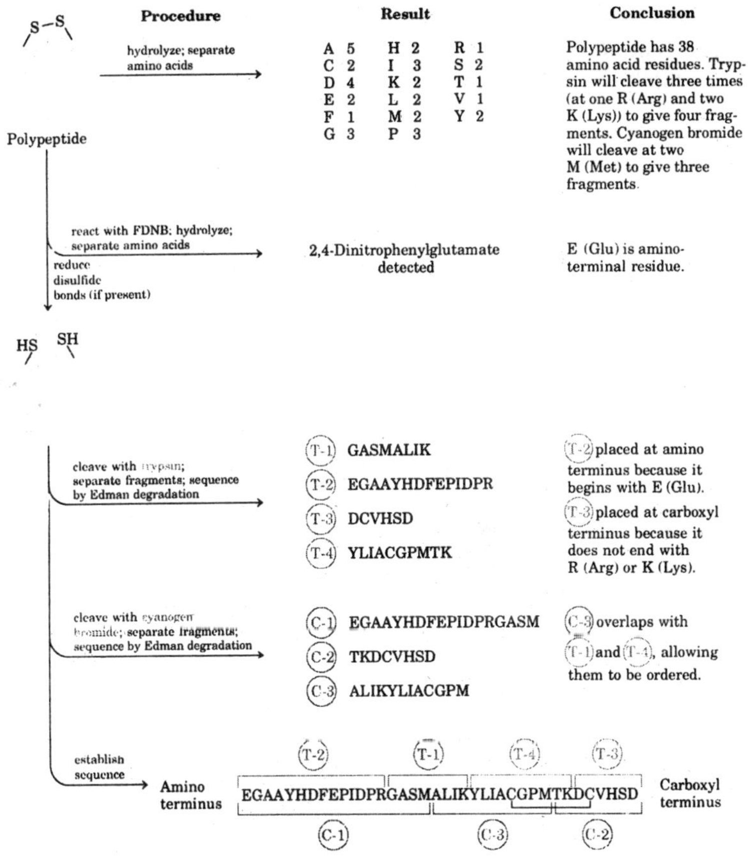

Figure 16. Cleaving proteins and sequencing and ordering the peptide fragments

permit the sequencing of short polypeptides (20 to 30 amino acid residues) in just a few minutes. In addition, with the development of rapid DNA sequencing methods, the elucidation of the genetic code, and the development of techniques for isolating genes, researchers can deduce the sequence of a polypeptide by determining the sequence of nucleotides in the gene that codes for it. The techniques used to determine protein and DNA sequences are complementary.

When the gene is available, sequencing the DNA can be faster and more accurate than sequencing the protein. Most proteins are now sequenced in this indirect way. If the gene has not been isolated, direct sequencing of peptides is necessary, and this can provide information (the location of disulfide bonds, for example) not available in a DNA sequence. In addition, a knowledge of the amino acid sequence of even a part of a polypeptide can greatly facilitate the isolation of the corresponding gene.

The array of methods now available to analyze both proteins and nucleic acids is ushering in a new disci-pline of "whole cell biochemistry." The complete sequence of an organism's DNA, its genome, is now available for organisms ranging from viruses to bacteria to multicellular eukaryotes. Genes are being discovered by the millions, including many that encode proteins with no known function.

To describe the entire protein complement encoded by an organism's DNA, researchers have coined the term proteome. The new disciplines of genomics and proteomics are complementing work carried out on cellular intermediary metabolism and nucleic acid metabolism to provide a new and increasingly complete picture of biochemistry at the level of cells and even organisms.

Chemically Synthesized Small Peptides and Proteins

Many peptides are potentially useful as pharmacologic agents, and their production is of considerable commercial importance.

There are three ways to obtain a peptide: (1) purification from tissue, a task often made difficult by the vanishingly low concentrations of some peptides; (2) genetic engineering; or (3) direct chemical synthesis. Powerful techniques now make direct chemical synthesis an attractive option in many cases. In addition to commercial applications, the synthesis of specific peptide portions of larger proteins is an increasingly important tool for the study of protein structure and function.

The complexity of proteins makes the traditional synthetic approaches of organic chemistry impractical for peptides with more than four or five amino acid residues. One problem is the difficulty of purifying the product after each step. The major breakthrough in this technology was provided by R. Bruce Merrifield in 1962. His innovation involved synthesizing a peptide while keeping it attached at one end to a solid support. The support is an insoluble polymer (resin) contained within a column, similar to that used for chromatographic procedures. The peptide is built up on this support one amino acid at a time using a standard set of reactions in a repeating cycle. At each successive step in the cycle, protective chemical groups block unwanted reactions.

The technology for chemical peptide synthesis is now automated. As in the sequencing reactions already considered, the most important limitation of the process is the efficiency of each chemical cycle, as can be seen by calculating the overall yields of peptides of various lengths when the yield for addition of each new amino acid is 96.0% versus 99.8%. Incomplete reaction at one stage can lead to formation of an impurity (in the form of a shorter peptide) in the next.

The chemistry has been optimized to permit the synthesis of proteins of 100 amino acid residues in a few days in reasonable yield. A very similar approach is used to synthesize nucleic acids. It is worth noting that this technology, impressive as it is, still pales when compared with biological processes. The same 100-amino-acid protein would be synthesized with exquisite fidelity in about 5 seconds in a bacterial cell.

A variety of new methods for the efficient ligation (joining together) of peptides has made possible the assembly of synthetic peptides into larger proteins. With these methods, novel forms of proteins can be created with precisely positioned chemical groups, including those that might not normally be found in a cellular protein. These novel forms provide new ways to test theories of enzyme catalysis, to create proteins with new chemical properties, and to design protein sequences that will fold into particular structures. This last application provides the ultimate test of our increasing ability to relate the primary structure of a peptide to the three-dimensional structure that it takes up in solution.

Amino Acid Sequences Provide Important Biochemical Information

Knowledge of the sequence of amino acids in a protein can offer insights into its three-dimensional structure and its function, cellular location, and evolution. Most of these insights are derived by searching for similarities with other known sequences. Thousands of sequences are known and available in databases accessible through the Internet. A comparison of a newly obtained sequence with this large bank of stored sequences often reveals relationships both surprising and enlightening.

Exactly how the amino acid sequence determines three-dimensional structure is not understood in detail, nor can we always predict function from sequence. However, protein families that have some shared structural or functional features can be readily identified on the basis of amino acid sequence similarities. Individual proteins are assigned to families based on the degree of similarity in amino acid sequence. Members of a family are usually identical across 25% or more of their sequences, and proteins in these families generally share at least some structural and functional characteristics.

Some families are defined, however, by identities involving only a few amino acid residues that are critical to a certain function. A number of similar substructures occur in many functionally unrelated proteins. These domains often fold into structural configurations that have an unusual degree of stability or that are specialized for a certain environment.

Evolutionary relationships can also be inferred from the structural and functional similarities within protein families. Certain amino acid sequences serve as signals that determine the cellular location, chemical modification, and half-life of a protein. Special signal sequences, usually at the amino terminus, are used to target certain proteins for export from the cell; other proteins are targeted for distribution to the nucleus, the cell surface, the cytosol, and other cellular locations.

Other sequences act as attachment sites for prosthetic groups, such as sugar groups in glycoproteins and lipids in lipoproteins. Some of these signals are well characterized and are easily recognized in the sequence of a newly characterized protein.

Protein Sequences and Evolution

The simple string of letters denoting the amino acid sequence of a given protein belies the wealth of information this sequence holds. As more protein sequences have become available, the development of more powerful methods for extracting information from them has become a major biochemical enterprise. Each protein's function relies on its three-dimensional structure, which in turn is determined largely by its primary structure. Thus, the biochemical information conveyed by a protein sequence is in principle limited only by our own understanding of structural and functional principles. On a different level of inquiry, protein sequences are beginning to tell us how the proteins evolved and, ultimately, how life evolved on this planet.

The field of molecular evolution is often traced to Emile Zuckerkandl and Linus Pauling, whose work in the mid-1960s advanced the use of nucleotide and protein sequences to explore evolution. The premise is deceptively straightforward. If two organisms are closely related, the sequences of their genes and proteins should be similar. The sequences increasingly diverge as the evolutionary distance between two organisms increases. The promise of this approach began to be realized in the 1970s, when Carl Woese used ribosomal RNA sequences to define archaebacteria as a group of living organisms distinct from other bacteria and eukaryotes.

Protein sequences offer an opportunity to greatly refine the available information. With the advent of genome projects investigating organisms from bacteria to humans, the number of available sequences is growing at an enormous rate. This information can be used to trace biological history. The challenge is in learning to read the genetic hieroglyphics. Evolution has not taken a simple linear path. Complexities abound in any attempt to mine the evolutionary information stored in protein sequences. For a given protein, the amino acid residues essential for the activity of the protein are conserved over evolutionary time. The residues that are less important to function may vary over time-that is, one amino acid may substitute for another-and these variable residues can provide the information used to trace evolution.

Amino acid substitutions are not always random, however. At some positions in the primary structure, the need to maintain protein function may mean that only particular amino acid substitutions can be tolerated. Some proteins have more variable amino acid residues than others. For these and other reasons, proteins can evolve at different rates. Another complicating factor in tracing evolutionary history is the rare transfer of a gene or group of genes from one organism to another, a process called lateral gene transfer.

The transferred genes may be quite similar to the genes they were derived from in the original organism, whereas most other genes in the same two organisms may be quite distantly related. An example of lateral gene transfer is the recent rapid spread of antibiotic-resistance genes in bacterial populations. The proteins derived from these transferred genes would not be good candidates for the study of bacterial evolution, because they share only a very limited evolutionary history with their "host" organisms.

In most cases, the families chosen for analysis have essential functions in cellular metabolism that must have been present in the earliest viable cells, thus greatly reducing the chance that they were introduced relatively recently by lateral gene transfer. For example, a protein called EF-1α is involved in the synthesis of proteins in all eukaryotes. A similar protein, EF-Tu, with the same function, is found in bacteria.

Similarities in sequence and function indicate that EF-1α and EF-Tu are members of a family of proteins that share a common ancestor. The members of protein families are called homologous proteins, or homologs. The concept of a homolog can be further refined. If two proteins within a family (that is, two homologs) are present in the same species, they are referred to as paralogs. Homologs from different species are called orthologs. The process of tracing evolution involves first identifying suitable families of homologous proteins and then using them to reconstruct evolutionary paths.

Homologs are identified using increasingly powerful computer programs that can directly compare two or more chosen protein sequences, or can search vast databases to find the evolutionary relatives of one selected protein sequence. The electronic search process can be thought of as sliding one sequence past the other until a section with a good match is found. Within this sequence alignment, a positive score is assigned for each position where the amino acid residues in the two sequences are identical-the value of the score varying from one program to the next-to provide a measure of the quality of the alignment. The process has some complications.

Sometimes the proteins being compared match well at, say, two sequence segments, and these segments are connected by less related sequences of different lengths. Thus the two matching segments cannot be aligned at the same time. To handle this, the computer program introduces "gaps" in one of the sequences to bring the matching segments into register.

Of course, if a sufficient number of gaps are introduced, almost any two sequences could be brought into some sort of alignment. To avoid uninformative alignments, the programs include penalties for each gap introduced, thus lowering the overall alignment score. With electronic trial and error, the program selects the alignment with the optimal score that maximizes identical amino acid residues while minimizing the introduction of gaps.

Identical amino acids are often inadequate to identify related proteins or, more importantly, to determine how closely related the proteins are on an evolutionary time scale. A more useful analysis includes a consideration of the chemical properties of substituted amino acids. When amino acid substitutions are found within a protein family, many of the differences may be conservative -that is, an amino acid residue is replaced by a residue having similar chemical properties.

For example, a Glu residue may substitute in one family member for the Asp residue found in another; both amino acids are negatively charged. Such a conservative substitution should logically garner a higher score in a sequence alignment than does a nonconservative substitution, such as the replacement of the Asp residue with a hydrophobic Phe residue.

To determine what scores to assign to the many different amino acid substitutions, Steven Henikoff and Jorja Henikoff examined the aligned sequences from a variety of different proteins. They did not analyze entire protein sequences, focusing instead on thousands of short conserved blocks where the fraction of identical amino acids was high and the alignments were thus reliable. Looking at the aligned sequence blocks, the Henikoffs analyzed the nonidentical amino acid residues within the blocks. Higher scores were given to nonidentical residues that occurred frequently than to those that appeared rarely.

For most efforts to find homologies and explore evolutionary relationships, protein sequences (derived either directly from protein sequencing or from the sequencing of the DNA encoding the protein) are superior to nongenic nucleic acid sequences (those that do not encode a protein or functional RNA). For a nucleic acid, with its four different types of residues, random alignment of nonhomologous sequences will generally yield matches for at least 25% of the positions.

Introduction of a few gaps can often increase the fraction of matched residues to 40% or more, and the probability of chance alignment of unrelated sequences becomes quite high. The 20 different amino acid residues in proteins greatly lower the probability of uninformative chance alignments of this type.

The programs used to generate a sequence alignment are complemented by methods that test the reliability of the alignments. A common computerized test is to shuffle the amino acid sequence of one of the proteins being compared to produce a random sequence, then instruct the program to align the shuffled sequence with the other, unshuffled one.

Scores are assigned to the new alignment, and the shuffling and alignment process is repeated many times. The original alignment, before shuffling, should have a score significantly higher than any of those within the distribution of scores generated by the random alignments; this increases the confidence that the sequence alignment has identified a pair of homologs. Note that the absence of a significant alignment score does not necessarily mean that no evolutionary relationship exists between two proteins.

Using a protein family to explore evolution requires the identification of family members with similar molecular functions in the widest possible range of organisms. Information from the family can then be used to trace the evolution of those organisms. By analyzing the sequence divergence in selected protein families, investigators can segregate organisms into classes based on their evolutionary relationships. This information must be reconciled with more classical examinations of the physiology and biochemistry of the organisms.

Certain segments of a protein sequence may be found in the organisms of one taxonomic group but not in other groups; these segments can be used as signature sequences for the group in which they are found. An example of a signature sequence is an insertion of 12 amino acids near the amino terminus of the EF-1α/EF-Tu proteins in all archaebacteria and eukaryotes but not in other types of bacteria. The signature is one of many biochemical clues that can help establish the evolutionary relatedness of eukaryotes and archaebacteria.

For example, the major taxa of bacteria can be distinguished by signature sequences in several different proteins. The β and γ proteobacteria have signature sequences in the Hsp70 and DNA gyrase protein families (families of proteins involved in protein folding and DNA replication, respectively) that are not present in any other bacteria, including the other proteobacteria.

The other types of proteobacteria (α, γ, ε), along with the β and γ proteobacteria, have a separate Hsp70 signature sequence and a signature in alanyl-tRNA synthetase that are not present in other bacteria. The appearance of unique signatures in the β and γ proteobacteria suggests the α, δ, and ε proteobacteria arose before their β and γ cousins. By considering the entire sequence of a protein, researchers can now construct more elaborate evolutionary trees with many species in each taxonomic group.

Figure 17 presents one such tree for bacteria, based on sequence divergence in the protein GroEL (a protein present in all bacteria that assists in the proper folding of proteins). The tree can be refined by basing it on the sequences of multiple proteins and by supplementing the sequence information with data on the unique biochemical and physiological properties of each species. There are many methods for generating trees, each with its own advantages and shortcomings, and many ways to represent the resulting evolutionary relationships

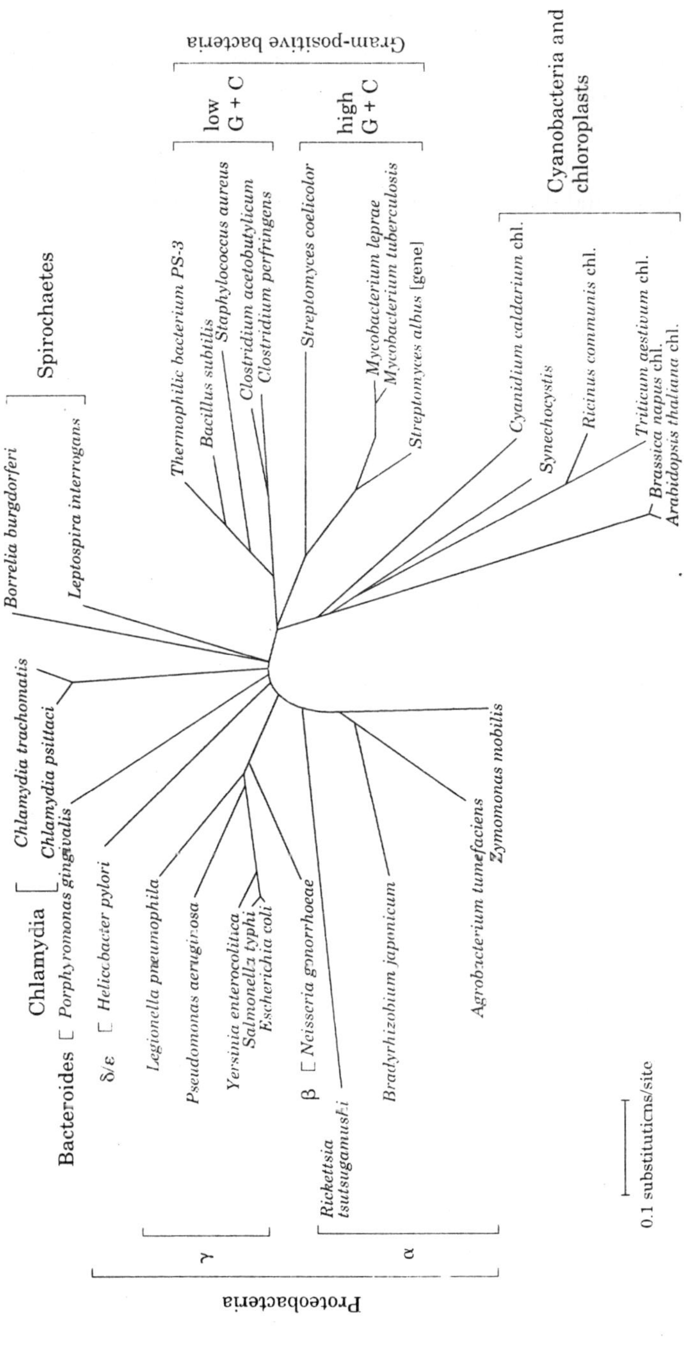

Figure 17. Evolutionary tree derived from amino acid sequence comparisons.

In Figure 17, the free end points of lines are called "external nodes"; each represents an extant species, and each is so labeled. The points where two lines come together, the "internal nodes," represent extinct ancestor species. In most representations, the lengths of the lines connecting the nodes are proportional to the number of amino acid substitutions separating one species from another.

5

Functions of Proteins

The spatial arrangement of atoms in a protein is called its conformation. The possible conformations of a protein include any structural state that can be achieved without breaking covalent bonds. A change in conformation could occur, for example, by rotation about single bonds. Of the numerous conformations that are theoretically possible in a protein containing hundreds of single bonds, one or a few generally predominate under biological conditions. The need for multiple stable conformations reflects the changes that must occur in most proteins as they bind to other molecules or catalyze reactions. The conformations existing under a given set of conditions are usually the ones that are thermodynamically the most stable, having the lowest Gibbs free energy (G). Proteins in any of their functional, folded conformations are called native proteins.

Protein's Confrontation

In the context of protein structure, the term stability can be defined as the tendency to maintain a native conformation. Native proteins are only marginally stable; the ΔG separating the folded and unfolded states in typical proteins under physiological conditions is in the range of only 20 to 65 kJ/mol. A given polypeptide chain can theoretically assume countless different conformations, and as a result the unfolded state of a protein is characterized by a high degree of conformational entropy.

This entropy, and the hydrogen-bonding interactions of many groups in the polypeptide chain with solvent (water), tend to maintain the unfolded state. The chemical interactions that counteract these effects and stabilize the native conformation include disulfide bonds and the weak (noncovalent) interactions: hydrogen bonds, and hydrophobic and ionic interactions. An appreciation of the role of these weak interactions is especially important to our understanding of how polypeptide chains fold into specific secondary and tertiary structures, and how they combine with other polypeptides to form quaternary structures.

About 200 to 460 kJ/mol are required to break a single covalent bond, whereas weak interactions can be disrupted by a mere 4 to 30 kJ/mol. Individual covalent bonds that contribute to the native conformations of proteins, such as disulfide bonds linking separate parts of a single polypeptide chain, are clearly much stronger than individual weak interactions. Yet, because they are so numerous, it is weak interactions that predominate as a stabilizing force in protein structure. In general, the protein conformation with the lowest free energy is the one with the maximum number of weak interactions.

The stability of a protein is not simply the sum of the free energies of formation of the many weak interactions within it. Every hydrogen-bonding group in a folded polypeptide chain was hydrogen-bonded to water prior to folding, and for every hydrogen bond formed in a protein, a hydrogen bond (of similar strength) between the same group and water was broken. The net stability contributed by a given weak interaction, or the *difference* in free energies of the folded and unfolded states, may be close to zero. We must therefore look elsewhere to explain why the native conformation of a protein is favored.

We find that the contribution of weak interactions to protein stability can be understood in terms of the properties of water.

Pure water contains a network of hydrogen-bonded H2O molecules. No other molecule has the hydrogen-bonding potential of water, and other molecules present in an aqueous solution disrupt the hydrogen bonding of water. When water surrounds a hydrophobic molecule, the optimal arrangement of hydrogen bonds results in a highly structured shell, or solvation layer, of water in the immediate vicinity.

The increased order of the water molecules in the solvation layer correlates with an unfavorable decrease in the entropy of the water. However, when nonpolar groups are clustered together, there is a decrease in the extent of the solvation layer because each group no longer presents its entire surface to the solution. The result is a favorable increase in entropy. This entropy term is the major thermodynamic driving force for the association of hydrophobic groups in aqueous solution. Hydrophobic amino acid side chains therefore tend to be clustered in a protein's interior, away from water.

Under physiological conditions, the formation of hydrogen bonds and ionic interactions in a protein is driven largely by this same entropic effect. Polar groups can generally form hydrogen bonds with water and hence are soluble in water. However, the number of hydrogen bonds per unit mass is generally greater for pure water than for any other liquid or solution, and there are limits to the solubility of even the most polar molecules as their presence causes a net decrease in hydrogen bonding per unit mass. Therefore, a solvation shell of structured water will also form to some extent around polar molecules.

Even though the energy of formation of an intramolecular hydrogen bond or ionic interaction between two polar groups in a macromolecule is largely canceled out by the elimination of such interactions between the same groups and water, the release of structured water when the intramolecular interaction is formed provides an entropic driving force for folding.

Most of the net change in free energy that occurs when weak interactions are formed within a protein is therefore derived from the increased entropy in the surrounding aqueous solution resulting from the burial of hydrophobic surfaces. This more than counterbalances the large loss of conformational entropy as a polypeptide is constrained into a single folded conformation.

Hydrophobic interactions are clearly important in stabilizing a protein conformation; the interior of a protein is generally a densely packed core of hydrophobic amino acid side chains. It is also important that any polar or charged groups in the protein interior have suitable partners for hydrogen bonding or ionic interactions.

One hydrogen bond seems to contribute little to the stability of a native structure, but the presence of hydrogen-bonding or charged groups without partners in the hydrophobic core of a protein can be so *destabilizing* that conformations containing these groups are often thermodynamically untenable. The favorable freeenergy change realized by combining such a group with a partner in the surrounding solution can be greater than the difference in free energy between the folded and unfolded states. In addition, hydrogen bonds between groups in proteins form cooperatively. Formation of one hydrogen bond facilitates the formation of additional hydrogen bonds.

The overall contribution of hydrogen bonds and other noncovalent interactions to the stabilization of protein conformation is still being evaluated. The interaction of oppositely charged groups that form an ion pair (salt bridge) may also have a stabilizing effect on one or more native conformations of some proteins. Most of the structural patterns outlined here reflect two simple rules:

(1) hydrophobic residues are largely buried in the protein interior, away from water; and

(2) the number of hydrogen bonds within the protein is maximized.

Insoluble proteins and proteins within membranes follow somewhat different rules because of their function or their environment, but weak interactions are still critical structural elements.

Peptide Bond

Protein Architecture—Primary Structure

Covalent bonds also place important constraints on the conformation of a polypeptide. In the late 1930s, Linus Pauling and Robert Corey embarked on a series of studies that laid the foundation for our present understanding of protein structure. They began with a careful analysis of the peptide bond. The α carbons of adjacent amino acid residues are separated by three covalent bonds, arranged as C_α—C—N—C_α.

X-ray diffraction studies of crystals of aminc acids and of simple dipeptides and tripeptides demonstrated that the peptide C—N bond is somewhat shorter than the C—N bond in a simple amine and that the atoms associated with the peptide bond are coplanar. This indicated a resonance or partial sharing of two pairs of electrons between the carbonyl oxygen and the amide nitrogen. The oxygen has a partial negative charge and the nitrogen a partial positive charge, setting up a small electric dipole.

The six atoms of the peptide group lie in a single plane, with the oxygen atom of the carbonyl group and the hydrogen atom of the amide nitrogen trans to each other. From these findings Pauling and Corey concluded that the peptide C—N bonds are unable to rotate freely because of their partial double-bond character. Rotation is permitted about the N—C_α and the C_α—C bonds. The backbone of a polypeptide chain can thus be pictured as a series of rigid planes with consecutive planes sharing a common point of rotation at C_α. The rigid peptide bonds limit the range of conformations that can be assumed by a polypeptide chain.

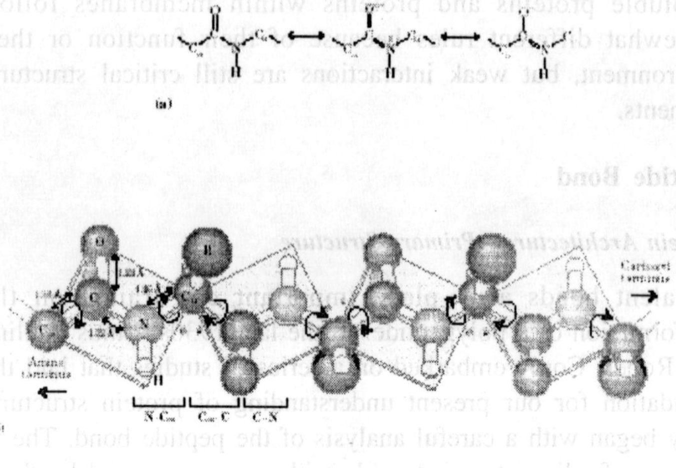

Figure.1. The planar peptide group. (a) Each peptide bond has some double-bond character due to resonance and cannot rotate. (b) Three bonds separate sequential α carbons in a polypeptide chain. The N—C_α and C_α—C bonds can rotate, with bond angles designated ϕ and ψ, respectively. (c) Both ϕ and ψ are defined as 0° when the two peptide bonds flanking that α–carbon are in the same plane and positioned as shown.

By convention, the bond angles resulting from rotations at C_α are labeled ϕ (phi) for the N—C_α bond and ψ (psi) for the C_α—C bond. Again by convention, both ϕ and ψ are defined as 180° when the polypeptide is in its fully extended conformation and all peptide groups are in the same plane. In principle, ϕ and ψ an have any value between –180° and +180°, but many values are prohibited by steric interference between atoms in the polypeptide backbone and amino acid side chains. The conformation in which both ϕ and ψ are 0° (Fig. 1c) is prohibited for this reason; this conformation is used merely as a reference point for describing the angles of rotation. Allowed values for ϕ and ψ are graphically revealed when ψ is plotted versus ϕ in a Ramachandran plot (Fig. 2), introduced by G. N. Ramachandran.

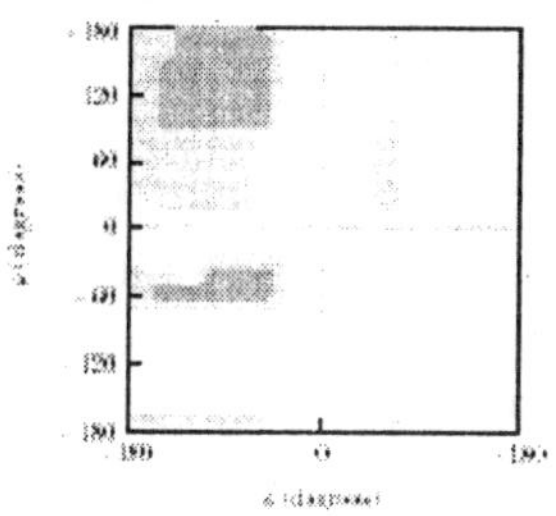

Figure.2. Ramachandran plot for L-Alaresidues

Protein Secondary Structure

The term secondary structure refers to the local conformation of some part of a polypeptide. The discussion of secondary structure most usefully focuses on common regular folding patterns of the polypeptide backbone. A few types of secondary structure are particularly stable and occur widely in proteins. The most prominent are the α helix and β conformations described below. Using fundamental chemical principles and a few experimental observations, Pauling and Corey predicted the existence of these secondary structures in 1951, several years before the first complete protein structure was elucidated.

The α Helix

Pauling and Corey were aware of the importance of hydrogen bonds in orienting polar chemical groups such as the CPO and NOH groups of the peptide bond. They also had the experimental results of William Astbury, who in the 1930s had conducted pioneering x-ray studies of proteins. Astbury demonstrated that the protein that makes up hair and porcupine quills (the fibrous protein α-keratin) has a regular structure that repeats every 5.15 to 5.2 Å. (The angstrom, Å, named after the physicist Anders J. Ångström, is equal to 0.1 nm. Although not an SI unit, it is used universally by structural biologists to describe atomic distances.)

With this information and their data on the peptide bond, and with the help of precisely constructed models, Pauling and Corey set out to determine the likely conformations of protein molecules. The simplest arrangement the polypeptide chain could assume with its rigid peptide bonds (but other single bonds free to rotate) is a helical structure, which Pauling and Corey called the α helix (Fig. 3). In this structure the polypeptide backbone is tightly wound around an imaginary axis drawn longitudinally through the middle of the helix, and the R groups of the amino acid residues protrude outward from the helical backbone.

The repeating unit is a single turn of the helix, which extends about 5.4 Å along the long axis, slightly greater than the periodicity Astbury observed on x-ray analysis of hair keratin. The amino acid residues in an α helix have conformations with $\psi = -45°$ to $-50°$ and $\phi = -60°$, and each helical turn includes 3.6 amino acid residues. The helical twist of the α helix found in all proteins is right-handed. The α helix proved to be the predominant structure in α-keratins. More generally, about one-fourth of all amino acid residues in polypeptides are found in α helices, the exact fraction varying greatly from one protein to the next.

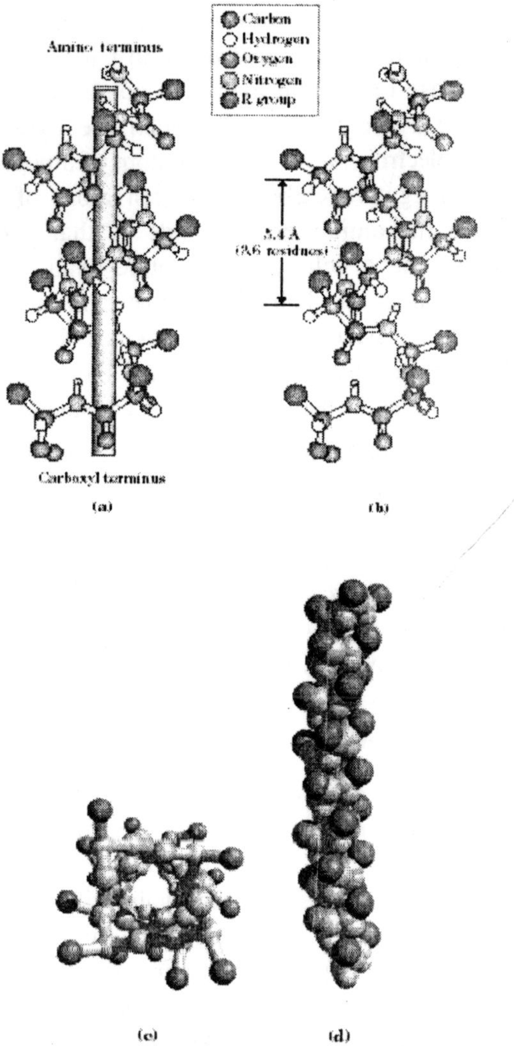

Figure.3. Four models of the a helix, showing different aspects of its structure. (a) Formation of a right-handed α helix. (b) Ball-and-stick model of a righthanded α helix, showing the intrachain hydrogen bonds. (c) The a helix as viewed from one end, looking down the longitudinal axis.As the space-filling model (d) shows, the atoms in the center of the α helix are in very close contact

Why does the α helix form more readily than many other possible conformations? The answer is, in part, that an α helix makes optimal use of internal hydrogen bonds. The structure is stabilized by a hydrogen bond between the hydrogen atom attached to the electronegative nitrogen atom of a peptide linkage and the electronegative carbonyl oxygen atom of the fourth amino acid on the amino-terminal side of that peptide bond (Fig.3b). Within the α helix, every peptide bond (except those close to each end of the helix) participates in such hydrogen bonding. Each successive turn of the α helix is held to adjacent turns by three to four hydrogen bonds. All the hydrogen bonds combined give the entire helical structure considerable stability.

Further model-building experiments have shown that an α helix can form in polypeptides consisting of either L- or D-amino acids. However, all residues must be of one stereoisomeric series; a D-amino acid will disrupt a regular structure consisting of L-amino acids, and vice versa. Naturally occurring L-amino acids can form either right- or left-handed α helices, but extended lefthanded helices have not been observed in proteins.

Amino Acid Sequence Affects α Helix Stability

Not all polypeptides can form a stable α helix. Interactions between amino acid side chains can stabilize or destabilize this structure. For example, if a polypeptide chain has a long block of Glu residues, this segment of the chain will not form an α helix at pH 7.0. The negatively charged carboxyl groups of adjacent Glu residues repel each other so strongly that they prevent formation of the α helix. For the same reason, if there are many adjacent Lys and/or Arg residues, which have positively charged R groups at pH 7.0, they will also repel each other and prevent formation of the α helix. The bulk and shape of Asn, Ser, Thr, and Cys residues can also destabilize an α helix if they are close together in the chain.

The twist of an α helix ensures that critical interactions occur between an amino acid side chain and the side chain three (and sometimes four) residues away on either side of it (Fig. 4). Positively charged amino acids are often found three residues away from negatively charged amino acids, permitting the formation of an ion pair. Two aromatic amino acid residues are often similarly spaced, resulting in a hydrophobic interaction.

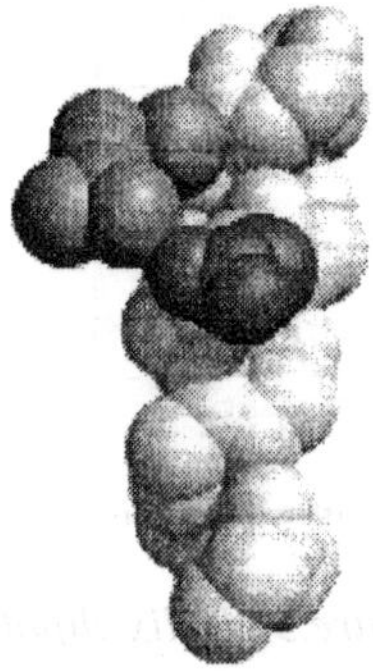

Figure.4. Interactions between R groups of amino acids three residues apart in an α helix

A constraint on the formation of the α helix is the presence of Pro or Gly residues. In proline, the nitrogen atom is part of a rigid ring, and rotation about the N—C_α bond is not possible. Thus, a Pro residue introduces a destabilizing kink in an α helix. In addition, the nitrogen atom of a Pro residue in peptide linkage has no substituent hydrogen to participate in hydrogen bonds with other residues. For these reasons, proline is only rarely found within an α helix. Glycine occurs infrequently in α helices for a different reason: it has more conformational flexibility than the other amino acid residues. Polymers of glycine tend to take up coiled structures quite different from an α helix. A final factor affecting the stability of an α helix in a polypeptide is the identity of the amino acid residues near the ends of the α-helical segment. A small electric dipole exists in each peptide bond. These dipoles are connected through the hydrogen bonds of the helix, resulting in a net dipole extending along the helix that increases with helix length (Fig.5).

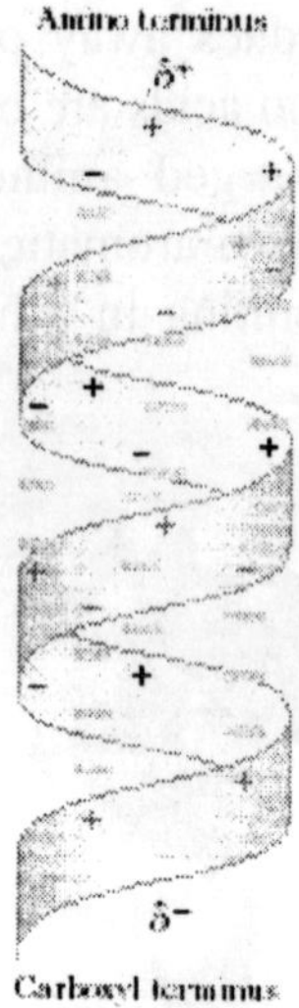

Figure.5. Helix dipole

The four amino acid residues at each end of the helix do not participate fully in the helix hydrogen bonds. The partial positive and negative charges of the helix dipole actually reside on the peptide amino and carbonyl groups near the amino-terminal and carboxyl-terminal ends of the helix, respectively. For this reason, negatively charged amino acids are often found near the amino terminus of the helical segment, where they have a stabilizing interaction with the positive charge of the helix dipole; a positively charged amino acid at the aminoterminal end is destabilizing. The opposite is true at the carboxyl-terminal end of the helical segment.

Thus, five different kinds of constraints affect the stability of an α helix:

(1) the electrostatic repulsion (or attraction) between successive amino acid residues with charged R groups,

(2) the bulkiness of adjacent R groups,

(3) the interactions between R groups spaced three (or four) residues apart,

(4) the occurrence of Pro and Gly residues, and

(5) the interaction between amino acid residues at the ends of the helical segment and the electric dipole inherent to the α helix.

The tendency of a given segment of a polypeptide chain to fold up as an α helix therefore depends on the identity and sequence of amino acid residues within the segment.

The β Conformation Organizes Polypeptide Chains into Sheets

Pauling and Corey predicted a second type of repetitive structure, the β conformation. This is a more extended conformation of polypeptide chains, and its structure has been confirmed by x-ray analysis. In the β conformation, the backbone of the polypeptide chain is extended into a zigzag rather than helical structure (Fig. 6).

The zigzag polypeptide chains can be arranged side by side to form a structure resembling a series of pleats. In this arrangement, called a β sheet, hydrogen bonds are formed between adjacent segments of polypeptide chain. The individual segments that form a β sheet are usually nearby on the polypeptide chain, but can also be quite distant from each other in the linear sequence of the polypeptide; they may even be segments in different polypeptide chains. The R groups of adjacent amino acids protrude from the zigzag structure in opposite directions, creating the alternating pattern seen in the side views in Figure 6.

The adjacent polypeptide chains in a β sheet can be either parallel or antiparallel (having the same or opposite amino-to-carboxyl orientations, respectively). The structures are somewhat similar, although the repeat period is shorter for the parallel conformation (6.5 Å, versus 7 Å for antiparallel) and the hydrogenbonding patterns are different.

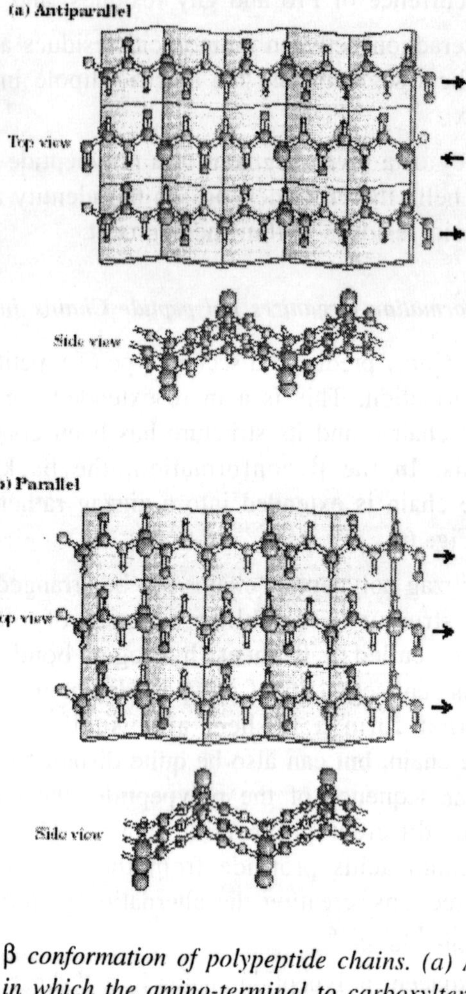

Figure.6. The β conformation of polypeptide chains. (a) Antiparallel β sheet, in which the amino-terminal to carboxylterminal orientation of adjacent chains is inverse. (b) Parallel β sheet.

Some protein structures limit the kinds of amino acids that can occur in the β sheet. When two or more β sheets are layered close together within a protein, the R groups of the amino acid residues on the touching surfaces must be relatively small. β-Keratins such as silk fibroin and the fibroin of spider webs have a very high content of Gly and Ala residues, the two amino acids with the smallest R groups. Indeed, in silk fibroin Gly and Ala alternate over large parts of the sequence.

β *Turns Are Common in Proteins*

In globular proteins, which have a compact folded structure, nearly one-third of the amino acid residues are in turns or loops where the polypeptide chain reverses direction (Fig.7). These are the connecting elements that link successive runs of α helix or β conformation. Particularly common are β turns that connect the ends of two adjacent segments of an antiparallel β sheet. The structure is a 180° turn involving four amino acid residues, with the carbonyl oxygen of the first residue forming a hydrogen bond with the amino-group hydrogen of the fourth. The peptide groups of the central two residues do not participate in any interresidue hydrogen bonding.

Gly and Pro residues often occur in β turns, the former because it is small and flexible, the latter because peptide bonds involving the imino nitrogen of proline readily assume the cis configuration (Fig. 7b), a form that is particularly amenable to a tight turn. Of the several types of β turns, the two shown in Figure 7a are the most common.

Beta turns are often found near the surface of a protein, where the peptide groups of the central two amino acid residues in the turn can hydrogen-bond with water. Considerably less common is the γ turn, a threeresidue turn with a hydrogen bond between the first and third residues.

Bond Angles and Amino Acid Content of Secondary Structures

The α helix and the β conformation are the major repetitive secondary structures in a wide variety of proteins, although other repetitive structures do exist in some specialized proteins. Every type of secondary structure can be completely described by the bond angles ϕ and ψ at each residue.

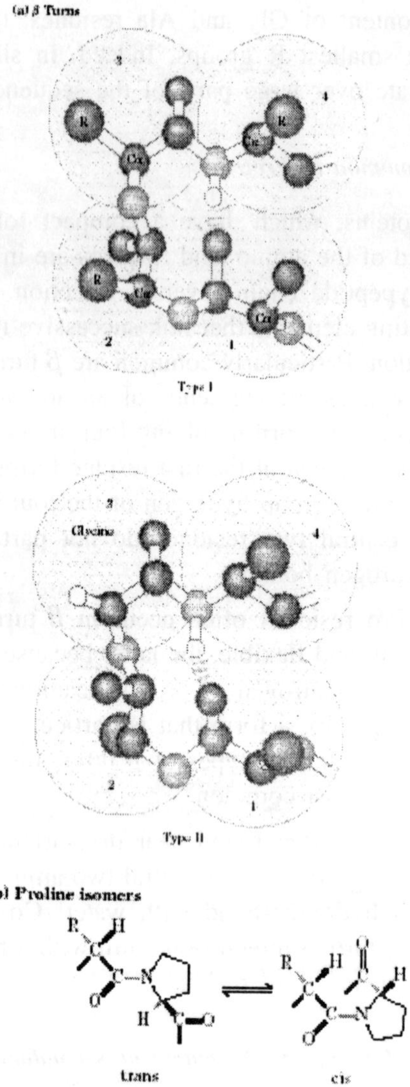

Figure.7. Structures of β turns. (a) Type I and type II β turns are most common; type I turns occur more than twice as frequently as type II. Type II β turns always have Gly as the third residue. (b) The trans and cis isomers of a peptide bond involving the imino nitrogen of proline.

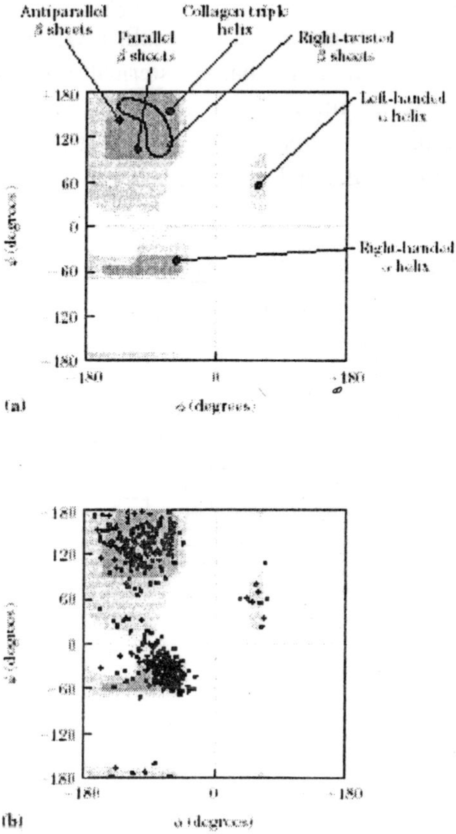

Figure.8. Ramachandran plots for a variety of structures. (a) The values of ϕ and ψ for various allowed secondary structures are overlaid on the plot from Fig.2. (b) The values of ϕ and ψ for all the amino acid residues except Gly in the enzyme pyruvate kinase are overlaid on the plot of theorctically allowed conformations.

As shown by a Ramachandran plot, the α helix and β conformation fall within a relatively restricted range of sterically allowed structures (Fig.8a). Most values of ϕ and ψ taken from known protein structures fall into the expected regions, with high concentrations near the α helix and β conformation values as predicted (Fig. 8b). The only amino acid residue often found in a conformation outside these regions is glycine. Because its side chain, a single hydrogen atom, is small, a Gly residue can take part in many conformations that are sterically forbidden for other amino acids. Some amino acids are accommodated better than others in the different types of secondary structures. An overall summary is presented in Figure 9.

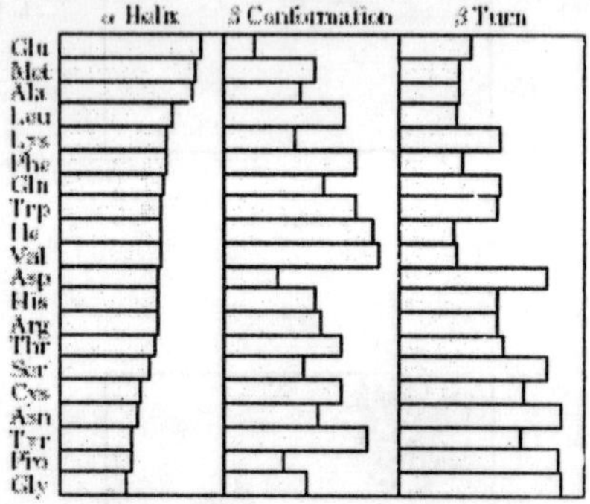

Figure.9. Relative probabilities that a given amino acid will occur in the three common types of secondary structure.

Some biases, such as the common presence of Pro and Gly residues in β turns and their relative absence in α helices, are readily explained by the known constraints on the different secondary structures. Other evident biases may be explained by taking into account the sizes or charges of side chains, but not all the trends shown in Figure 9 are understood.

Protein Tertiary and Quaternary Structures

The overall three-dimensional arrangement of all atoms in a protein is referred to as the protein's tertiary structure. Whereas the term "secondary structure" refers to the spatial arrangement of amino acid residues that are adjacent in the primary structure, tertiary structure includes longer-range aspects of amino acid sequence.

Amino acids that are far apart in the polypeptide sequence and that reside in different types of secondary structure may interact within the completely folded structure of a protein. The location of bends (including β turns) in the polypeptide chain and the direction and angle of these bends are determined by the number and location of specific bend-producing residues, such as Pro, Thr, Ser, and Gly. Interacting segments of polypeptide chains are held in their characteristic tertiary positions by different kinds of weak bonding interactions (and sometimes by covalent bonds such as disulfide cross-links) between the segments.

Some proteins contain two or more separate polypeptide chains, or subunits, which may be identical or different. The arrangement of these protein subunits in three-dimensional complexes constitutes quaternary structure.In considering these higher levels of structure, it is useful to classify proteins into two major groups: fibrous proteins, having polypeptide chains arranged in long strands or sheets, and globular proteins, having polypeptide chains folded into a spherical or globular shape.

The two groups are structurally distinct: fibrous proteins usually consist largely of a single type of secondary structure; globular proteins often contain several types of secondary structure. The two groups differ functionally in that the structures that provide support, shape, and external protection to vertebrates are made of fibrous proteins, whereas most enzymes and regulatory proteins are globular proteins. Certain fibrous proteins played a key role in the development of our modern understanding of protein structure and provide particularly clear examples of the relationship between structure and function.

Fibrous Proteins

α-Keratin, collagen, and silk fibroin nicely illustrate the relationship between protein structure and biological function. Fibrous proteins share properties that give strength and/or flexibility to the structures in which they occur. In each case, the fundamental structural unit is a simple repeating element of secondary structure. All fibrous proteins are insoluble in water, a property conferred by a high concentration of hydrophobic amino acid residues both in the interior of the protein and on its surface. These hydrophobic surfaces are largely buried by packing many similar polypeptide chains together to form elaborate supramolecular complexes.

Table 1. Secondary Structures and Properties of Fibrous Proteins

Structure	Characteristics	Examples of occurrence
α Helix, cross-linked by disulfide	Tough, insoluble protective structures of varying hardness and flexibility	α-Keratin of hair, feathers, and nails
β Conformation	Soft, flexible filaments	Silk fibroin
Collagen triple helix	High tensile strength, without stretch ,	Collagen of tendons bone matrix

α-Keratin

The α-keratins have evolved for strength. Found in mammals, these proteins constitute almost the entire dry weight of hair, wool, nails, claws, quills, horns, hooves, and much of the outer layer of skin. The α-keratins are part of a broader family of proteins called intermediate filament (IF) proteins. Other IF proteins are found in the cytoskeletons of animal cells. All IF proteins have a structural function and share structural features exemplified by the α-keratins.

The α-keratin helix is a right-handed α helix, the same helix found in many other proteins. Francis Crick and Linus Pauling in the early 1950s independently suggested that the α helices of keratin were arranged as a coiled coil. Two strands of α-keratin, oriented in parallel, are wrapped about each other to form a supertwisted coiled coil. The supertwisting amplifies the strength of the overall structure, just as strands are twisted to make a strong rope.

The twisting of the axis of an α helix to form a coiled coil explains the discrepancy between the 5.4 Å per turn predicted for an α helix by Pauling and Corey and the 5.15 to 5.2 Å repeating structure observed in the x-ray diffraction of hair. The helical path of the supertwists is left-handed, opposite in sense to the α helix. The surfaces where the two α helices touch are made up of hydrophobic amino acid residues, their R groups meshed together in a regular interlocking pattern. This permits a close packing of the polypeptide chains within the left-handed supertwist. Not surprisingly, α-keratin is rich in the hydrophobic residues Ala, Val, Leu, Ile, Met, and Phe.

An individual polypeptide in the α-keratin coiled coil has a relatively simple tertiary structure, dominated by an α-helical secondary structure with its helical axis twisted in a left-handed superhelix. The intertwining of the two α-helical polypeptides is an example of quaternary structure. Coiled coils of this type are common structural elements in filamentous proteins and in the muscle protein myosin. The quaternary structure of α-keratin can be quite complex. Many coiled coils can be assembled into large supramolecular complexes, such as the arrangement of α-keratin to form the intermediate filament of hair.

The strength of fibrous proteins is enhanced by covalent cross-links between polypeptide chains within the multihelical "ropes" and between adjacent chains in a supramolecular assembly. In α-keratins, the cross-links stabilizing quaternary structure are disulfide bonds. In the hardest and toughest α-keratins, such as those of rhinoceros horn, up to 18% of the residues are cysteines involved in disulfide bonds.

Collagen

Like the α-keratins, collagen has evolved to provide strength. It is found in connective tissue such as tendons, cartilage, the organic matrix of bone, and the cornea of the eye. The collagen helix is a unique secondary structure quite distinct from the α helix. It is left-handed and has three amino acid residues per turn. Collagen is also a coiled coil, but one with distinct tertiary and quaternary structures: three separate polypeptides, called α chains, are supertwisted about each other. The superhelical twisting is right-handed in collagen, opposite in sense to the left-handed helix of the α chains.

There are many types of vertebrate collagen. Typically they contain about 35% Gly, 11% Ala, and 21% Pro and 4-Hyp (4-hydroxyproline, an uncommon amino acid). The food product gelatin is derived from collagen; it has little nutritional value as a protein, because collagen is extremely low in many amino acids that are essential in the human diet. The unusual amino acid content of collagen is related to structural constraints unique to the collagen helix. The amino acid sequence in collagen is generally a repeating tripeptide unit, Gly–X–Y, where X is often Pro, and Y is often 4-Hyp. Only Gly residues can be accommodated at the very tight junctions between the individual α chains; The Pro and 4-Hyp residues permit the sharp twisting of the collagen helix. The amino acid sequence and the supertwisted quaternary structure of collagen allow a very close packing of its three polypeptides. 4-Hydroxyproline has a special role in the structure of collagen—and in human history.

Dehydrohydroxylysinonorleucine

The tight wrapping of the α chains in the collagen triple helix provides tensile strength greater than that of a steel wire of equal cross section. Collagen fibrils are supramolecular assemblies consisting of triple-helical collagen molecules (sometimes referred to as tropocollagen molecules) associated in a variety of ways to provide different degrees of tensile strength.

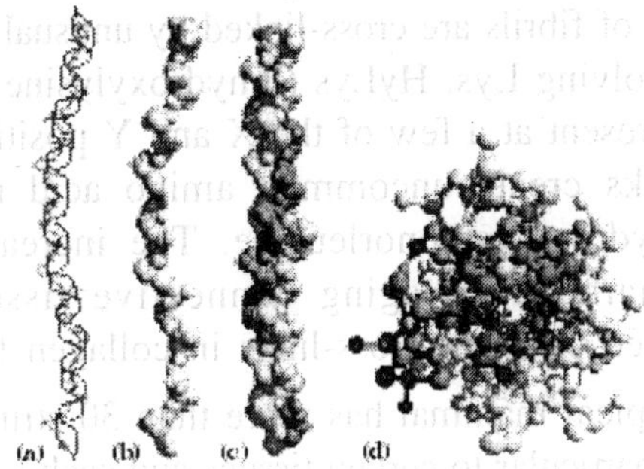

Figure.10. Structure of collagen. (a) The α chain of collagen has a repeating secondary structure unique to this protein. The repeating tripeptide sequence Gly–X–Pro or Gly–X–4-Hyp adopts a left-handed helical structure with three residues per turn. (b) Space-filling model of the same α chain. (c) Three of these helices wrap around one another with a right-handed twist. (d) The three-stranded collagen superhelix shown from one end, in a ball-and-stick representation.

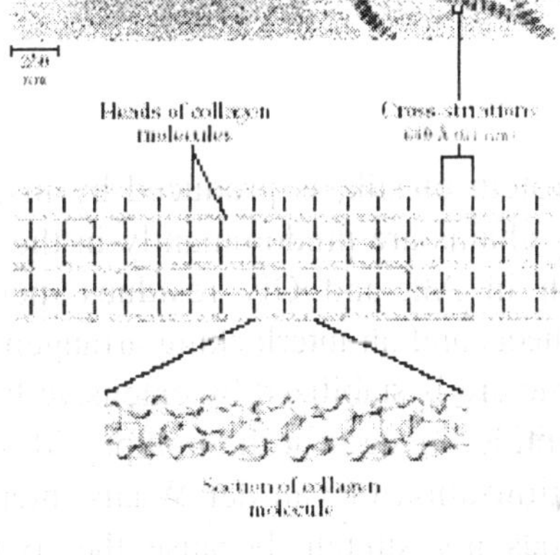

Figure. 11 Structure of collagen fibrils

The α chains of collagen molecules and the collagen molecules of fibrils are cross-linked by unusual types of covalent bonds involving Lys, HyLys (5-hydroxylysine), or His residues that are present at a few of the X and Y positions in collagens. These links create uncommon amino acid residues such as dehydrohydroxylysinonorleucine. The increasingly rigid and brittle character of aging connective tissue results from accumulated covalent cross-links in collagen fibrils.

A typical mammal has more than 30 structural variants of collagen, particular to certain tissues and each somewhat different in sequence and function. Some human genetic defects in collagen structure illustrate the close relationship between amino acid sequence and three-dimensional structure in this protein. Osteogenesis imperfecta is characterized by abnormal bone formation in babies; Ehlers-Danlos syndrome is characterized by loose joints. Both conditions can be lethal, and both result from the substitution of an amino acid residue with a larger R group (such as Cys or Ser) for a single Gly residue in each α chain (a different Gly residue in each disorder). These single-residue substitutions have a catastrophic effect on collagen function because they disrupt the Gly–X–Y repeat that gives collagen its unique helical structure. Given its role in the collagen triple helix, Gly cannot be replaced by another amino acid residue without substantial deleterious effects on collagen structure.

Silk Fibroin

Fibroin, the protein of silk, is produced by insects and spiders. Its polypeptide chains are predominantly in the β conformation. Fibroin is rich in Ala and Gly residues, permitting a close packing of β sheets and an interlocking arrangement of R groups. The overall structure is stabilized by extensive hydrogen bonding between all peptide linkages in the polypeptides of each β sheet and by the optimization of van der Waals interactions between sheets. Silk does not stretch, because the β conformation is already highly extended. However, the structure is flexible because the sheets are held together by numerous weak interactions rather than by covalent bonds such as the disulfide bonds in α-keratins.

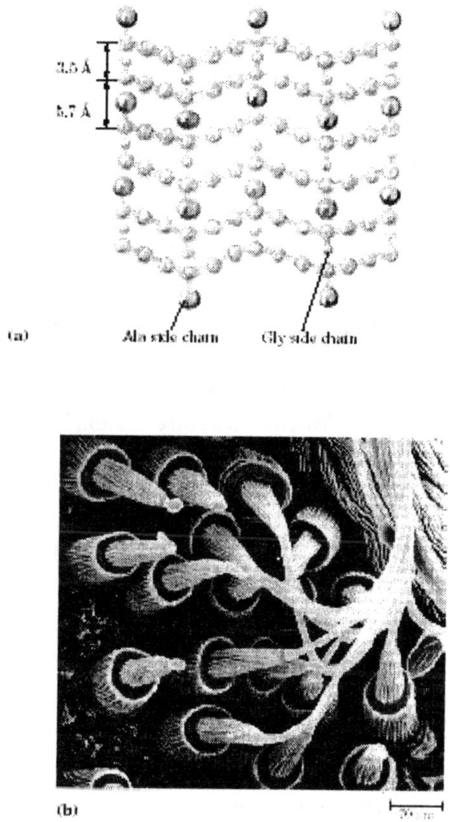

Figure.12. Structure of silk. (a) Fibroin consists of layers of antiparallel b *sheets rich in Ala (purple) and Gly (yellow) residues. (b) Strands of fibroin (blue) emerge from the spinnerets of a spider in this colorized electron micrograph.*

Structural Diversity in Globular Proteins

In a globular protein, different segments of a polypeptide chain (or multiple polypeptide chains) fold back on each other. As illustrated in Figure 13, this folding generates a compact form relative to polypeptides in a fully extended conformation. The folding also provides the structural diversity necessary for proteins to carry out a wide array of biological functions. Globular proteins include enzymes, transport proteins, motor proteins, regulatory proteins, immunoglobulins, and proteins with many other functions.

Figure. 13. Globular protein structures are compact and varied

Complexity of Globular Protein Structure Example of Myoglobin

The first breakthrough in understanding the three-dimensional structure of a globular protein came from x-ray diffraction studies of myoglobin carried out by John Kendrew and his colleagues in the 1950s. Myoglobin is a relatively small, oxygen-binding protein of muscle cells. It functions both to store oxygen and to facilitate oxygen diffusion in rapidly contracting muscle tissue.

Myoglobin contains a single polypeptide chain of 153 amino acid residues of known sequence and a single iron protoporphyrin, or heme, group. The same heme group is found in hemoglobin, the oxygen-binding protein of erythrocytes, and is responsible for the deep red-brown color of both myoglobin and hemoglobin. Myoglobin is particularly abun-dant in the muscles of diving mammals such as the whale, seal, and porpoise, whose muscles are so rich in this protein that they are brown. Storage and distribution of oxygen by muscle myoglobin permit these animals to remain submerged for long periods of time. Figure 14 shows several structural representations of myoglobin, illustrating how the polypeptide chain is folded in three dimensions—its tertiary structure.

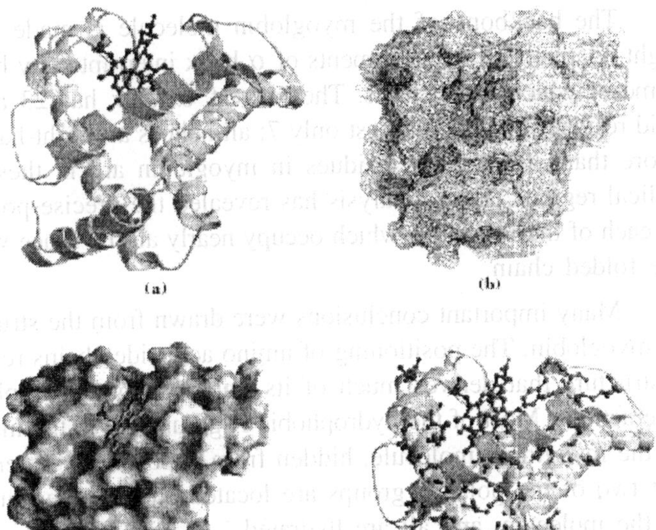

Figure.14. Tertiary structure of sperm whale myoglobin (a) The polypeptide backbone, shown in a ribbon representation of a type introduced by Jane Richardson, which highlights regions of secondary structure. The α-helical regions are evident. (b) A "mesh" image emphasizes the protein surface. (c) A surface contour image is useful for visualizing pockets in the protein where other molecules might bind. (d) A ribbon representation, including side chains for the hydrophobic residues Leu, Ile, Val, and Phe. (e) A space-filling model with all amino acid side chains.

The backbone of the myoglobin molecule is made up of eight relatively straight segments of α helix interrupted by bends, some of which are β turns. The longest α helix has 23 amino acid residues and the shortest only 7; all helices are right-handed. More than 70% of the residues in myoglobin are in these α-helical regions. X-ray analysis has revealed the precise position of each of the R groups, which occupy nearly all the space within the folded chain.

Many important conclusions were drawn from the structure of myoglobin. The positioning of amino acid side chains reflects a structure that derives much of its stability from hydrophobic interactions. Most of the hydrophobic R groups are in the interior of the myoglobin molecule, hidden from exposure to water. All but two of the polar R groups are located on the outer surface of the molecule, and all are hydrated.

The myoglobin molecule is so compact that its interior has room for only four molecules of water. This dense hydrophobic core is typical of globular proteins. The fraction of space occupied by atoms in an organic liquid is 0.4 to 0.6; in a typical crystal the fraction is 0.70 to 0.78, near the theoretical maximum. In a globular protein the fraction is about 0.75, comparable to that in a crystal. In this packed environment, weak interactions strengthen and reinforce each other. For example, the nonpolar side chains in the core are so close together that short-range van der Waals interactions make a significant contribution to stabilizing hydrophobic interactions.

Deduction of the structure of myoglobin confirmed some expectations and introduced some new elements of secondary structure. As predicted by Pauling and Corey, all the peptide bonds are in the planar trans configuration. The α helices in myoglobin provided the first direct experimental evidence for the existence of this type of secondary structure. Three of the four Pro residues of myoglobin are found at bends The fourth Pro residue occurs within an α helix, where it creates a kink necessary for tight helix packing. Other bends contain Ser, Thr, and Asn residues, which are among the amino acids whose bulk and shape tend to make them incompatible with α -helical structure if they are in close proximity in the amino acid sequence. The flat heme group rests in a crevice, or pocket, in the myoglobin molecule. The iron atom in the center of the heme group has two bonding (coordination) positions perpendicular to the plane of the heme (Fig.15).

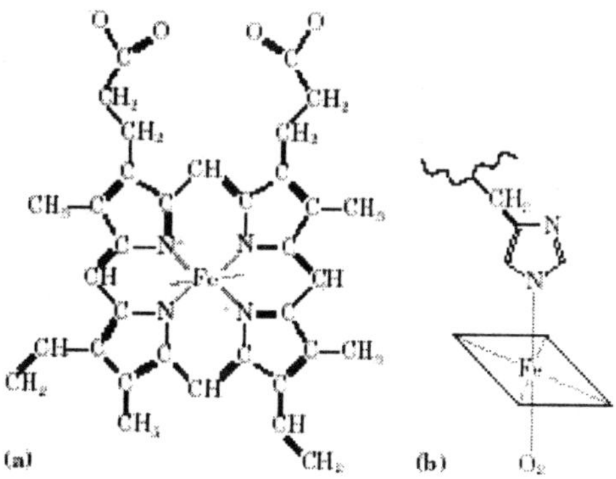

Figure.15. The heme group. (a) Heme consists of a complex organic ring structure, protoporphyrin, to which is bound an iron atom in its ferrous (Fe^{2+}) state. (b) In myoglobin and hemoglobin, one of the perpendicular coordination bonds is bound to a nitrogen atom of a His residue.

One of these is bound to the R group of the His residue at position 93; the other is the site at which an O_2 molecule binds. Within this pocket, the accessibility of the heme group to solvent is highly restricted. This is important for function, because free heme groups in an oxygenated solution are rapidly oxidized from the ferrous (Fe^{2+}) form, which is active in the reversible binding of O_2, to the ferric (Fe^{3+}) form, which does not bind O_2. Knowledge of the structure of myoglobin allowed researchers for the first time to understand in detail the correlation between the structure and function of a protein. Many different myoglobin structures have been elucidated, allowing investigators to see how the structure changes when oxygen or other molecules bind to it. Hundreds of proteins have been subjected to similar analysis since then. Today, techniques such as NMR spectroscopy supplement x-ray diffraction data, providing more information on a protein's structure.

Tertiary Structures of Globular Proteins

With elucidation of the tertiary structures of hundreds of other globular proteins by x-ray analysis, it became clear that myoglobin illustrates only one of many ways in which a polypeptide chain can be folded. In Figure 16 the structures of cytochrome *c,* lysozyme, and ribonuclease are compared. These proteins have different amino acid sequences and different tertiary structures, reflecting differences in function. All are relatively small and easy to work with, facilitating structural analysis.

Cytochrome *c* is a component of the respiratory chain of mitochondria. Like myoglobin, cytochrome *c* is a heme protein. It contains a single polypeptide chain of about 100 residues and a single heme group. In this case, the protoporphyrin of the heme group is covalently attached to the polypeptide. Only about 40% of the polypeptide is in α-helical segments, compared with 70% of the myoglobin chain. The rest of the cytochrome *c* chain contains β turns and irregularly coiled and extended segments.

Lysozyme is an enzyme abundant in egg white and human tears that catalyzes the hydrolytic cleavage of polysaccharides in the protective cell walls of some families of bacteria. Lysozyme, because it can lyse, or degrade, bacterial cell walls, serves as a bactericidal agent. As in cytochrome *c,* about 40% of its 129 amino acid residues are in α-helical segments, but the arrangement is different and some β-sheet structure is also present. Four disulfide bonds contribute stability to this structure. The α helices line a long crevice in the side of the molecule, called the active site, which is the site of substrate binding and catalysis.

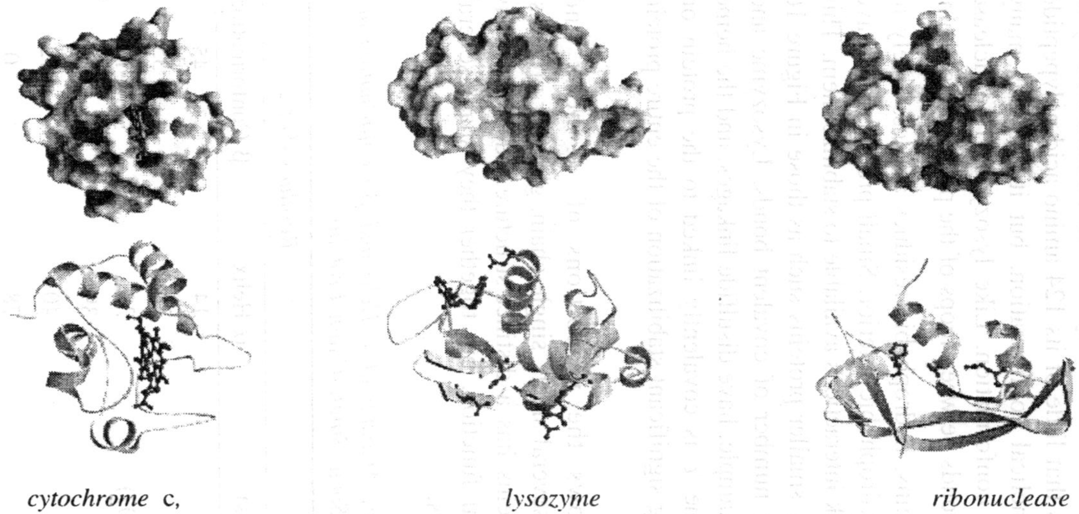

Figure.16. Three-dimensional structures of some small proteins. Shown here are cytochrome c, lysozyme and ribonuclease. Each protein is shown in surface contour and in a ribbon representation, in the same orientation. In the ribbon depictions, regions in the β conformation are represented by flat arrows and the α helices are represented by spiral ribbons. Key functional groups (the heme in cytochrome c; amino acid side chains in the active site of lysozyme and ribonuclease) are shown. Disulfide bonds are shown (in the ribbon representations).

Ribonuclease, another small globular protein, is an enzyme secreted by the pancreas into the small intestine, where it catalyzes the hydrolysis of certain bonds in the ribonucleic acids present in ingested food. Its tertiary structure, determined by x-ray analysis, shows that little of its 124 amino acid polypeptide chain is in an α-helical conformation, but it contains many segments in the β conformation. Like lysozyme, ribonuclease has four disulfide bonds between loops of the polypeptide chain.

In small proteins, hydrophobic residues are less likely to be sheltered in a hydrophobic interior. Small proteins also have fewer potential weak interactions available to stabilize them. This explains why many smaller proteins such as those in Figure 16 are stabilized by a number of covalent bonds. Lysozyme and ribonuclease, for example, have disulfide linkages, and the heme group in cytochrome *c* is covalently linked to the protein on two sides, providing significant stabilization of the entire protein structure.

Table 2 shows the proportions of α helix and β conformation in several small, single-chain, globular proteins. Each of these proteins has a distinct structure, adapted for its particular biological function, but together they share several important properties.

Table 2. Approximate Amounts of α Helix and β Conformation in Some Single-Chain Proteins

	Residues (%)	
Protein (total residues)	α Helix	β Conformation
Chymotrypsin (247)	14	45
Ribonuclease (124)	26	35
Carboxypeptidase (307)	38	17
Cytochrome *c* (104)	39	0
Lysozyme (129)	40	12
Myoglobin (153)	78	0

Each is folded compactly, and in each case the hydrophobic amino acid side chains are oriented toward the interior (away from water) and the hydrophilic side chains are on the surface. The structures are also stabilized by a multitude of hydrogen bonds and some ionic interactions.

Common Structural Patterns of Globular Proteins

For the beginning student, the very complex tertiary structures of globular proteins much larger than those shown in Figure 16 are best approached by focusing on structural patterns that recur in different and often unrelated proteins. The three-dimensional structure of a typical globular protein can be considered an assemblage of polypeptide segments in the α-helix and β-sheet conformations, linked by connecting segments. The structure can then be described to a first approximation by defining how these segments stack on one another and how the segments that connect them are arranged. This formalism has led to the development of databases that allow informative comparisons of protein structures, complementing other databases that permit comparisons of protein sequences.

An understanding of a complete three-dimensional structure is built upon an analysis of its parts. We begin by defining terms used to describe protein substructures, then turn to the folding rules elucidated from analysis of the structures of many proteins.

Supersecondary structures, also called motifs or simply folds, are particularly stable arrangements of several elements of secondary structure and the connections between them. There is no universal agreement among biochemists on the application of the three terms, and they are often used interchangeably. The terms are also applied to a wide range of structures. Recognized motifs range from simple to complex, sometimes appearing in repeating units or combinations. A single large motif may comprise the entire protein. We have already encountered one well-studied motif, the coiled coil of α-keratin, also found in a number of other proteins.

Polypeptides with more than a few hundred amino acid residues often fold into two or more stable, globular units called **domains.** In many cases, a domain from a large protein will retain its correct three-dimensional structure even when it is separated from the remainder of the polypeptide chain. A protein with multiple domains may appear to have a distinct globular lobe for each domain (Fig. 17), but, more commonly, extensive contacts between domains make individual domains hard to discern. Different domains often have distinct functions, such as the binding of small molecules or interaction with other proteins. Small proteins usually have only one domain.

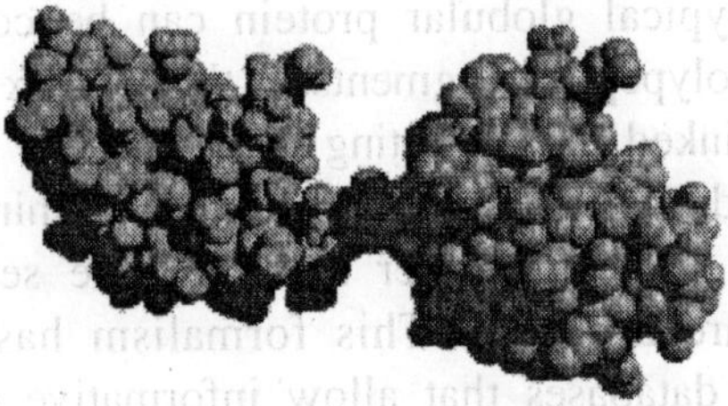

Figure.17. Structural domains in the polypeptide troponin C

Folding of polypeptides is subject to an array of physical and chemical constraints. A sampling of the prominent folding rules that have emerged provides an opportunity to introduce some simple motifs.

1. Hydrophobic interactions make a large contribution to the stability of protein structures. Burial of hydrophobic amino acid R groups so as to exclude water requires at least two layers of secondary structure. Two simple motifs, the β–α–β loop and the α–α corner (Fig.18a), create two layers.
2. Where they occur together in proteins, α helices and β sheets generally are found in different structural layers. This is because the backbone of a polypeptide segment in the β conformation cannot readily hydrogen-bond to an α helix aligned with it.

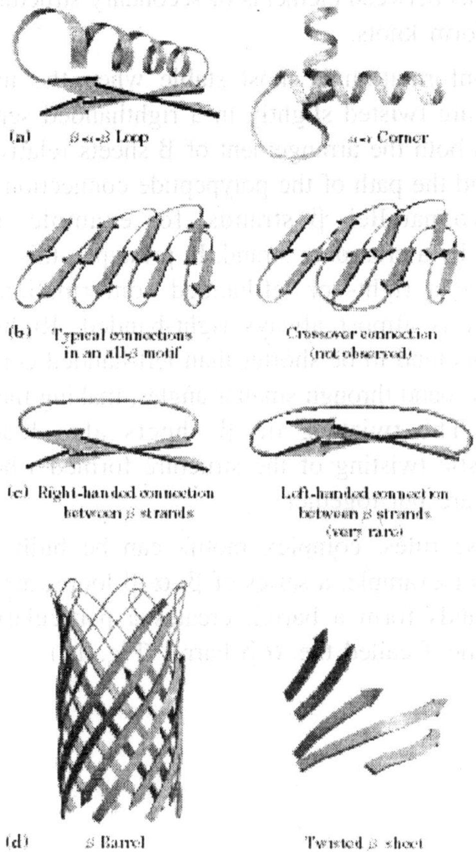

Figure.18. Stable folding patterns in proteins. (a) Two simple and common motifs that provide two layers of secondary structure. (b) Connections between β strands in layered β sheets. (c) Because of the twist in β strands, connections between strands are generally right-handed. (d) Two arrangements of β strands stabilized by the tendency of the strands to twist.

3. Polypeptide segments adjacent to each other in the primary sequence are usually stacked adjacent to each other in the folded structure. Although distant segments of a polypeptide may come together in the tertiary structure, this is not the norm.

4. Connections between elements of secondary structure cannot cross or form knots.
5. The β conformation is most stable when the individual segments are twisted slightly in a righthanded sense. This influences both the arrangement of β sheets relative to one another and the path of the polypeptide connection between them. Two parallel β strands, for example, must be connected by a crossover strand. In principle, this crossover could have a right- or lefthanded conformation, but in proteins it is almost always right-handed. Right-handed connections tend to be shorter than left-handed connections and tend to bend through smaller angles, making them easier to form. The twisting of β sheets also leads to a characteristic twisting of the structure formed when many segments are put together.

Following these rules, complex motifs can be built up from simple ones. For example, a series of β–α–β loops, arranged so that the β strands form a barrel, creates a particularly stable and common motif called the α/β barrel (Fig. 19).

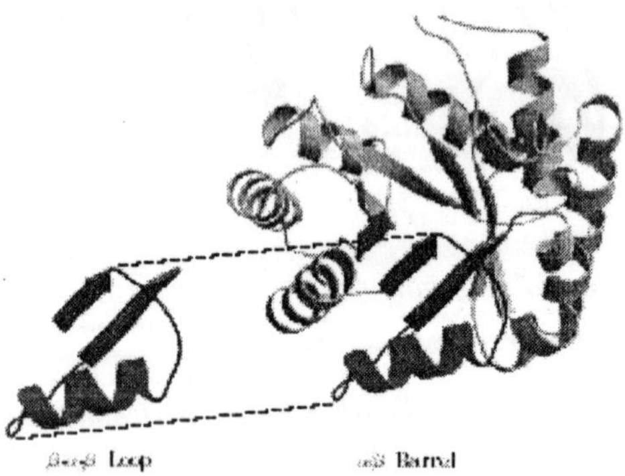

Figure.19. Constructing large motifs from smaller ones

In this structure, each parallel β segment is attached to its neighbor by an α-helical segment. All connections are right-handed. The α/β barrel is found in many enzymes, often with a binding site for a cofactor or substrate in the form of a pocket near one end of the barrel. Note that domains exhibiting similar folding patterns are said to have the same motif even though their constituent α helices and β sheets may differ in length.

Structural Classification of Proteins

The complexities of tertiary structure are decreased by considering substructures. Taking this idea further, researchers have organized the complete contents of databases according to hierarchical levels of structure. The Structural Classification of Proteins (SCOP) database offers a good example of this very important trend in biochemistry.

At the highest level of classification, the SCOP database borrows a scheme already in common use, in which protein structures are divided into four classes: all α, all β, α/β (in which the α and β segments are interspersed or alternate), and α + β (in which the α and β regions are somewhat segregated. Within each class are tens to hundreds of different folding arrangements, built up from increasingly identifiable substructures. Some of the substructure arrangements are very common, others have been found in just one protein.

Many examples of recurring domain or motif structures are available, and these reveal that protein tertiary structure is more reliably conserved than primary sequence. The comparison of protein structures can thus provide much information about evolution. Proteins with significant primary sequence similarity, and/or with demonstrably similar structure and function, are said to be in the same protein family.

A strong evolutionary relationship is usually evident within a protein family. For example, the globin family has many different proteins with both structural and sequence similarity to myoglobin. Two or more families with little primary sequence similarity sometimes make use of the same major structural motif and have functional similarities; these families are grouped as superfamilies. An evolutionary relationship between the families in a superfamily is considered probable, even though time and functional distinctions —hence different adaptive pressures—may have erased many of the telltale sequence relationships.

A protein family may be widespread in all three domains of cellular life, the Bacteria, Archaea, and Eukarya, suggesting a very ancient origin. Other families may be present in only a small group of organisms, indicating that the structure arose more recently. Tracing the natural history of structural motifs, using structural classifications in databases such as SCOP, provides a powerful complement to sequence analyses in tracing many evolutionary relationships.

The SCOP database is curated manually, with the objective of placing proteins in the correct evolutionary framework based on conserved structural features. Two similar enterprises, the CATH (*c*lass, *a*rchitecture, *t*opology, and *h*omologous superfamily) and FSSP (fold classification based on *s*tructure-*s*tructure alignment of *p*roteins) databases, make use of more automated methods and can provide additional information. Structural motifs become especially important in defining protein families and superfamilies.

Improved classification and comparison systems for proteins lead inevitably to the elucidation of new functional relationships. Given the central role of proteins in living systems, these structural comparisons can help illuminate every aspect of biochemistry, from the evolution of individual proteins to the evolutionary history of complete metabolic pathways.

Protein Quaternary Structures

Many proteins have multiple polypeptide subunits. The association of polypeptide chains can serve a variety of functions. Many multisubunit proteins have regulatory roles; the binding of small molecules may affect the interaction between subunits, causing large changes in the protein's activity in response to small changes in the concentration of substrate or regulatory molecules.

In other cases, separate subunits can take on separate but related functions, such as catalysis and regulation. Some associations, such as the fibrous proteins considered earlier and the coat proteins of viruses, serve primarily structural roles. Some very large protein assemblies are the site of complex, multistep reactions. One example is the ribosome, site of protein synthesis, which incorporates dozens of protein subunits along with a number of RNA molecules.

A multisubunit protein is also referred to as a multimer. Multimeric proteins can have from two to hundreds of subunits. A multimer with just a few subunits is often called an oligomer. If a multimer is composed of a number of nonidentical subunits, the overall structure of the protein can be asymmetric and quite complicated. However, most multimers have identical subunits or repeating groups of nonidentical subunits, usually in symmetric arrangements. The repeating structural unit in such a multimeric protein, whether it is a single subunit or a group of subunits, is called a protomer.

The first oligomeric protein for which the threedimensional structure was determined was hemoglobin, which contains four polypeptide chains and four heme prosthetic groups, in which the iron atoms are in the ferrous (Fe^{2+}) state. The protein portion, called globin, consists of two α chains (141 residues each) and two β chains (146 residues each).

Note that in this case α and β do not refer to secondary structures. Because hemoglobin is four times as large as myoglobin, much more time and effort were required to solve its three-dimensional structure by x-ray analysis, finally achieved by Max Perutz, John Kendrew, and their colleagues in 1959. The subunits of hemoglobin are arranged in symmetric pairs, each pair having one α and one β subunit. Hemoglobin can therefore be described either as a tetramer or as a dimer of $\alpha\beta$ protomers.

Identical subunits of multimeric proteins are generally arranged in one or a limited set of symmetric patterns. A description of the structure of these proteins requires an understanding of conventions used to define symmetries. Oligomers can have either rotational symmetry or helical symmetry; that is, individual subunits can be superimposed on others by rotation about one or more rotational axes, or by a helical rotation. In proteins with rotational symmetry, the subunits pack about the rotational axes to form closed structures. Proteins with helical symme-try tend to form structures that are more open-ended, with subunits added in a spiraling array.

There are several forms of rotational symmetry. The simplest is cyclic symmetry, involving rotation about a single axis (Fig.20a). If subunits can be superimposed by rotation about a single axis, the protein has a symmetry defined by convention as C*n* (C for cyclic, *n* for the number of subunits related by the axis). The axis itself is described as an *n*-fold rotational axis. The αβ protomers of hemoglobin are related by C_2 symmetry.

A somewhat more complicated rotational symmetry is dihedral symmetry, in which a twofold rotational axis intersects an *n*-fold axis at right angles. The symmetry is defined as D*n* (Fig.20b). A protein with dihedral symmetry has 2*n* protomers. Proteins with cyclic or dihedral symmetry are particularly common. More complex rotational symmetries are possible, but only a few are regularly encountered. One example is icosahedral symmetry. An icosahedron is a regular 12-cornered polyhedron having 20 equilateral triangular faces (Fig. 20c).

Each face can be brought to coincidence with another by rotation about one or more of three rotational axes. This is a common structure in virus coats, or capsids. The human poliovirus has an icosahedral capsid. Each triangular face is made up of three protomers, each protomer containing single copies of four different polypeptide chains, three of which are accessible at the outer surface. Sixty protomers form the 20 faces of the icosahedral shell enclosing the genetic material (RNA).

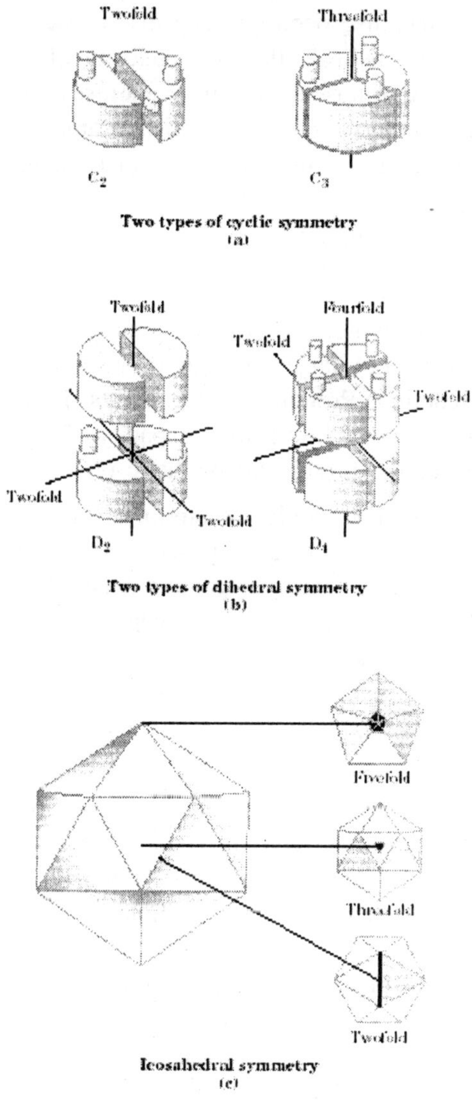

Figure.20. Rotational symmetry in proteins. (a) Cyclic symmetry, (b) Dihedral symmetry (c) Icosahedral symmetry.

The other major type of symmetry found in oligomers, helical symmetry, also occurs in capsids. Tobacco mosaic virus is a right-handed helical filament made up of 2,130 identical subunits. This cylindrical structure encloses the viral RNA. Proteins with subunits arranged in helical filaments can also form long, fibrous structures such as the actin filaments of muscle.

Limits to the Size of Proteins

The relatively large size of proteins reflects their functions. The function of an enzyme, for example, requires a stable structure containing a pocket large enough to bind its substrate and catalyze a reaction. Protein size has limits, however, imposed by two factors: the genetic coding capacity of nucleic acids and the accuracy of the protein biosynthetic process.

The use of many copies of one or a few proteins to make a large enclosing structure (capsid) is important for viruses because this strategy conserves genetic material. Remember that there is a linear correspondence between the sequence of a gene in the nucleic acid and the amino acid sequence of the protein for which it codes.

The nucleic acids of viruses are much too small to encode the information required for a protein shell made of a single polypeptide. By using many copies of much smaller polypeptides, a much shorter nucleic acid is needed for coding the capsid subunits, and this nucleic acid can be efficiently used over and over again. Cells also use large complexes of polypeptides in muscle, cilia, the cytoskeleton, and other structures. It is simply more efficient to make many copies of a small polypeptide than one copy of a very large protein. In fact, most proteins with a molecular weight greater than 100,000 have multiple subunits, identical or different.

The second factor limiting the size of proteins is the error frequency during protein biosynthesis. The error frequency is low (about 1 mistake per 10,000 amino acid residues added), but even this low rate results in a high probability of a damaged protein if the protein is very large. Simply put, the potential for incorporating a "wrong" amino acid in a protein is greater for a large protein than for a small one.

Protein Denaturation and Folding

All proteins begin their existence on a ribosome as a linear sequence of amino acid residues. This polypeptide must fold during and following synthesis to take up its native conformation. We have seen that a native protein conformation is only marginally stable. Modest changes in the protein's environment can bring about structural changes that can affect function. We now explore the transition that occurs between the folded and unfolded states.

Loss of Protein Structure

Protein structures have evolved to function in particular cellular environments. Conditions different from those in the cell can result in protein structural changes, large and small. A loss of three-dimensional structure sufficient to cause loss of function is called denaturation. The denatured state does not necessarily equate with complete unfolding of the protein and randomization of conformation. Under most conditions, denatured proteins exist in a set of partially folded states that are poorly understood.

Most proteins can be denatured by heat, which affects the weak interactions in a protein (primarily hydrogen bonds) in a complex manner. If the temperature is increased slowly, a protein's conformation generally remains intact until an abrupt loss of structure (and function) occurs over a narrow temperature range (Fig. 21). The abruptness of the change suggests that unfolding is a cooperative process: loss of structure in one part of the protein destabilizes other parts. The effects of heat on proteins are not readily predictable. The very heat-stable proteins of thermophilic bacteria have evolved to function at the temperature of hot springs (~100 °C). Yet the structures of these proteins often differ only slightly from those of homologous proteins derived from bacteria such as *Escherichia coli.* How these small differences promote structural stability at high temperatures is not yet understood.

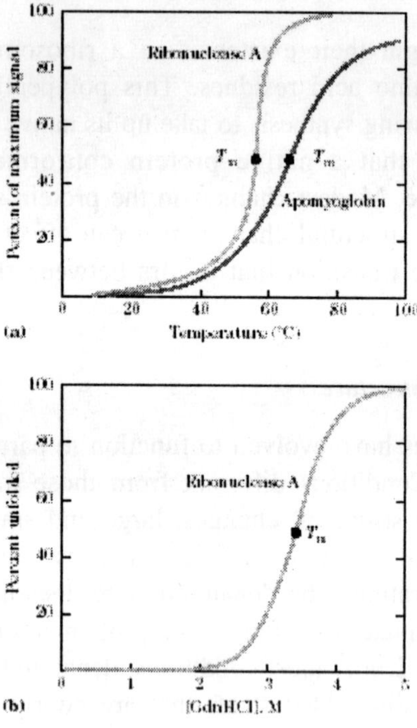

Figure. 21. Protein denaturation. (a) Thermal denaturation of horse apomyoglobin and ribonuclease A. (b) Denaturation of disulfide-intact ribonuclease A by guanidine hydrochloride (GdnHCl), monitored by circular dichroism.

Proteins can be denatured not only by heat but by extremes of pH, by certain miscible organic solvents such as alcohol or acetone, by certain solutes such as urea and guanidine hydrochloride, or by detergents. Each of these denaturing agents represents a relatively mild treatment in the sense that no covalent bonds in the polypeptide chain are broken. Organic solvents, urea, and detergents act primarily by disrupting the hydrophobic interactions that make up the stable core of globular proteins; extremes of pH alter the net charge on the protein, causing electrostatic repulsion and the disruption of some hydrogen bonding. The denatured states obtained with these various treatments need not be equivalent.

Amino Acid Sequence and Tertiary Structure

The tertiary structure of a globular protein is determined by its amino acid sequence. The most important proof of this came from experiments showing that denaturation of some proteins is reversible. Certain globular proteins denatured by heat, extremes of pH, or denaturing reagents will regain their native structure and their biological activity if returned to conditions in which the native conformation is stable. This process is called renaturation.

A classic example is the denaturation and renaturation of ribonuclease. Purified ribonuclease can be completely denatured by exposure to a concentrated urea solution in the presence of a reducing agent. The reducing agent cleaves the four disulfide bonds to yield eight Cys residues, and the urea disrupts the stabilizing hydrophobic interactions, thus freeing the entire polypeptide from its folded conformation.

Denaturation of ribonuclease is accompanied by a complete loss of catalytic activity. When the urea and the reducing agent are removed, the randomly coiled, denatured ribonuclease spontaneously refolds into its correct tertiary structure, with full restoration of its catalytic activity.

The refolding of ribonuclease is so accurate that the four intrachain disulfide bonds are re-formed in the same positions in the renatured molecule as in the native ribonuclease.As calculated mathematically, the eight Cys residues could recombine at random to form up to four disulfide bonds in 105 different ways. In fact, an essentially random distribution of disulfide bonds is obtained when the disulfides are allowed to reform in the presence of denaturant, indicating that weak bonding interactions are required for correct positioning of disulfide bonds and assumption of the native conformation.

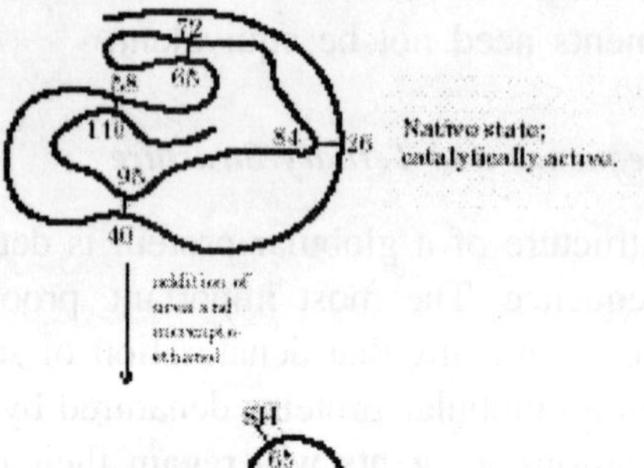

Figure.22. Renaturation of unfolded, denatured ribonuclease

This classic experiment, carried out by Christian Anfinsen in the 1950s, provided the first evidence that the amino acid sequence of a polypeptide chain contains all the information required to fold the chain into its native, three-dimensional structure. Later, similar results were obtained using chemically synthesized, catalytically active ribonuclease. This eliminated the possibility that some minor contaminant in Anfinsen's purified ribonuclease preparation might have contributed to the renaturation of the enzyme, thus dispelling any remaining doubt that this enzyme folds spontaneously.

Polypeptides Fold Rapidly by a Stepwise Process

In living cells, proteins are assembled from amino acids at a very high rate. For example, *E. coli* cells can make a complete, biologically active protein molecule containing 100 amino acid residues in about 5 seconds at 37°C. How does such a polypeptide chain arrive at its native conformation?

Let's assume conservatively that each of the amino acid residues could take up 10 different conformations on average, giving 10^{100} different conformations for the polypeptide. Let's also assume that the protein folds itself spontaneously by a random process in which it tries out all possible conformations around every single bond in its backbone until it finds its native, biologically active form.

If each conformation were sampled in the shortest possible time (~10^{-13} second, or the time required for a single molecular vibration), it would take about 10^{77} years to sample all possible conformations. Thus protein folding cannot be a completely random, trial-and-error process. There must be shortcuts. This problem was first pointed out by Cyrus Levinthal in 1968 and is sometimes called Levinthal's paradox.

The folding pathway of a large polypeptide chain is unquestionably complicated, and not all the principles that guide the process have been worked out. However, extensive study has led to the development of several plausible models. In one, the folding process is envisioned as hierarchical. Local secondary structures form first. Certain amino acid sequences fold readily into α helices or β sheets, guided by constraints we have reviewed in our discussion of secondary structure. This is followed by longer-range interactions between, say, two α helices that come together to form stable supersecondary structures. The process continues until complete domains form and the entire polypeptide is folded.

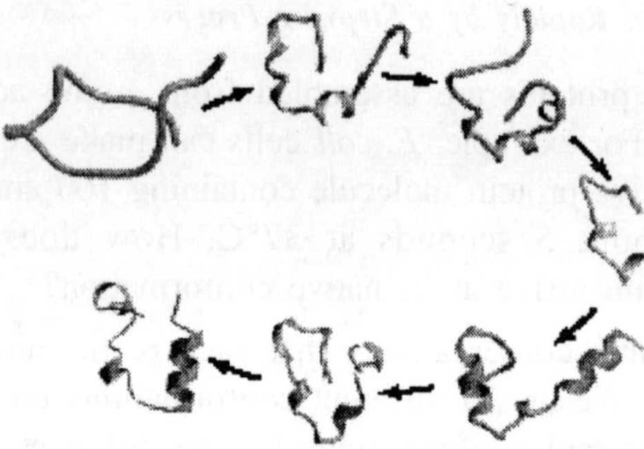

Figure.23. A simulated folding pathway

In an alternative model, folding is initiated by a spontaneous collapse of the polypeptide into a compact state, mediated by hydrophobic interactions among nonpolar residues. The state resulting from this "hydrophobic collapse" may have a high content of secondary structure, but many amino acid side chains are not entirely fixed.

The collapsed state is often referred to as a molten globule. Most proteins probably fold by a process that incorporates features of both models. Instead of following a single pathway, a population of peptide molecules may take a variety of routes to the same end point, with the number of different partly folded conformational species decreasing as folding nears completion.

Thermodynamically, the folding process can be viewed as a kind of free-energy funnel (Fig.23). The unfolded states are characterized by a high degree of conformational entropy and relatively high free energy. As folding proceeds, the narrowing of the funnel repre-sents a decrease in the number of conformational species present. Small depressions along the sides of the free-energy funnel represent semistable intermediates that can briefly slow the folding process. At the bottom of the funnel, an ensemble of folding intermediates has been reduced to a single native conformation (or one of a small set of native conformations).

Defects in protein folding may be the molecular basis for a wide range of human genetic disorders. For example, cystic fibrosis is caused by defects in a membrane-bound protein called *c*ystic *f*ibrosis *t*ransmembrane conductance *r*egulator (CFTR), which acts as a channel for chloride ions. The most common cystic fibrosis–causing mutation is the deletion of a Phe residue at position 508 in CFTR, which causes improper protein folding. Many of the diseaserelated mutations in collagen also cause defective folding. An improved understanding of protein folding may lead to new therapies for these and many other diseases.

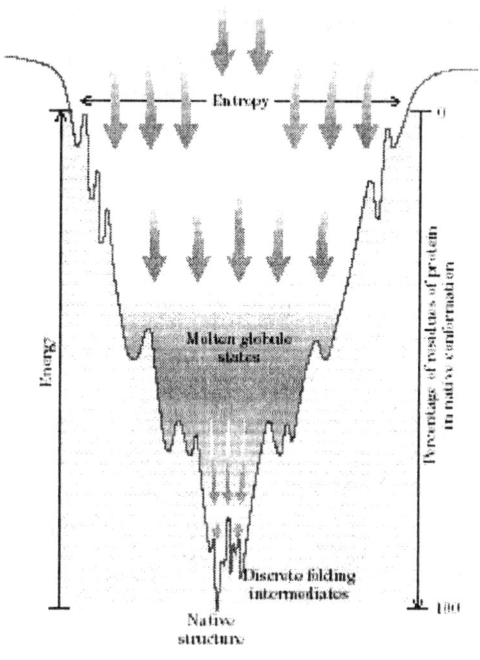

Figure.24. The thermodynamics of protein folding depicted as a free-energy funnel

Thermodynamic stability is not evenly distributed over the structure of a protein—the molecule has regions of high and low stability. For example, a protein may have two stable domains joined by a segment with lower structural stability, or one small part of a domain may have a lower stability than the remainder. The regions of low stability allow a protein to alter its conformation between two or more states.

Assisted Folding

Not all proteins fold spontaneously as they are synthesized in the cell. Folding for many proteins is facilitated by the action of specialized proteins. Molecular chaperones are proteins that interact with partially folded or improperly folded polypeptides, facilitating correct folding pathways or providing microenvironments in which folding can occur. Two classes of molecular chaperones have been well studied. Both are found in organisms ranging from bacteria to humans. The first class, a family of proteins called Hsp70, generally have a molecular weight near 70,000 and are more abundant in cells stressed by elevated temperatures. Hsp70 proteins bind to regions of unfolded polypeptides that are rich in hydrophobic residues, preventing inappropriate aggregation. These chaperones thus "protect" proteins that have been denatured by heat and peptides that are being synthesized. Hsp70 proteins also block the folding of certain proteins that must remain unfolded until they have been translocated across membranes.

Some chaperones also facilitate the quaternary assembly of oligomeric proteins. The Hsp70 proteins bind to and release polypeptides in a cycle that also involves several other proteins and ATP hydrolysis. Figure 25 illustrates chaperoneassisted folding as elucidated for the chaperones DnaK and DnaJ in *E. coli,* homologs of the eukaryotic Hsp70 and Hsp40. DnaK and DnaJ were first identified as proteins required for in vitro replication of certain viral DNA molecules. The second class of chaperones is called chaperonins.

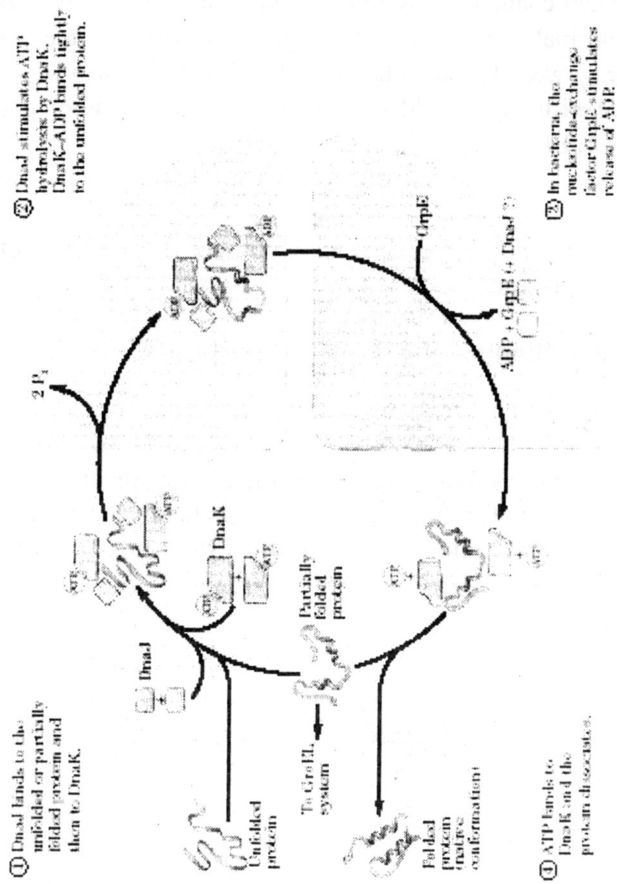

Figure.25. Chaperones in protein folding

The second class of chaperones is called chaperonins. These are elaborate protein complexes required for the folding of a number of cellular proteins that do not fold spontaneously. In *E. coli* an estimated 10% to 15% of cellular proteins require the resident chaperonin system, called GroEL/GroES, for folding under normal conditions. These proteins first became known when they were found to be necessary for the growth of certain bacterial viruses. Unfolded proteins are bound within pockets in the GroEL complex, and the pockets are capped transiently by the GroES "lid". GroEL undergoes undergoes substantial conformational changes, coupled to ATP hydrolysis and the binding and release of GroES, which promote folding of the bound polypeptide.

Reversible Binding of a Protein to a Ligand

The functions of many proteins involve the reversible binding of other molecules. A molecule bound reversibly by a protein is called a ligand. A ligand may be any kind of molecule, including another protein. The transient nature of proteinligand interactions is critical to life, allowing an organism to respond rapidly and reversibly to changing environmental and metabolic circumstances.

A ligand binds at a site on the protein called the binding site, which is complementary to the ligand in size, shape, charge, and hydrophobic or hydrophilic character. Furthermore, the interaction is specific: the protein can discriminate among the thousands of different molecules in its environment and selectively bind only one or a few. A given protein may have separate binding sites for several different ligands. These specific molecular interactions are crucial in maintaining the high degree of order in a living system.

Proteins are flexible. Changes in conformation may be subtle, reflecting molecular vibrations and small movements of amino acid residues throughout the protein. A protein flexing in this way is sometimes said to "breathe." Changes in conformation may also be quite dramatic, with major segments of the protein structure moving as much as several nanometers. Specific conformational changes are frequently essential to a protein's function.

The binding of a protein and ligand is often coupled to a conformational change in the protein that makes the binding site more complementary to the ligand, permitting tighter binding. The structural adaptation that occurs between protein and ligand is called induced fit. In a multisubunit protein, a conformational change in one subunit often affects the conformation of other subunits.

Interactions between ligands and proteins may be regulated, usually through specific interactions with one or more additional ligands. These other ligands may cause conformational changes in the protein that affect the binding of the first ligand. Enzymes represent a special case of protein function. Enzymes bind and chemically transform other molecules -they catalyze reactions. The molecules acted upon by enzymes are called reaction substrates rather than ligands, and the ligand-binding site is called the catalytic site or active site.

Myoglobin and hemoglobin may be the most-studied and best-understood proteins. They were the first proteins for which three-dimensional structures were determined, and our current understanding of myoglobin and hemoglobin is garnered from the work of thousands of biochemists over several decades. Most important, these molecules illustrate almost every aspect of that most central of biochemical processes: the reversible binding of a ligand to a protein. This classic model of protein function tells us a great deal about how proteins work.

Oxygen Can Be Bound to a Heme Prosthetic Group

Oxygen is poorly soluble in aqueous solutions and cannot be carried to tissues in sufficient quantity if it is simply dissolved in blood serum. Diffusion of oxygen through tissues is also ineffective over distances greater than a few millimeters. The evolution of larger, multicellular animals depended on the evolution of proteins that could transport and store oxygen. However, none of the amino acid side chains in proteins is suited for the reversible binding of oxygen molecules. This role is filled by certain transition metals, among them iron and copper, that have a strong tendency to bind oxygen.

Multicellular organisms exploit the properties of metals, most commonly iron, for oxygen transport. However, free iron promotes the formation of highly reactive oxygen species such as hydroxyl radicals that can damage DNA and other macromolecules. Iron used in cells is therefore bound in forms that sequester it and/or make it less reactive. In multicellular organisms-especially those in which iron, in its oxygen-carrying capacity, must be transported over large distances-iron is often incorporated into a protein-bound prosthetic group called heme.

Heme (or haem) consists of a complex organic ring structure, protoporphyrin, to which is bound a single iron atom in its ferrous (Fe^{2+}) state. The iron atom has six coordination bonds, four to nitrogen atoms that are part of the flat porphyrin ring system and two perpendicular to the porphyrin. The coordinated nitrogen atoms (which have an electron-donating character) help prevent conversion of the heme iron to the ferric (Fe^{3+}) state. Iron in the Fe^{2+} state binds oxygen reversibly; in the Fe3 state it does not bind oxygen. Heme is found in a number of oxygen-transporting proteins, as well as in some proteins, such as the cytochromes, that participate in oxidation-reduction (electron-transfer) reactions.

In free heme molecules (heme not bound to protein), reaction of oxygen at one of the two "open" coordination bonds of iron (perpendicular to the plane of the porphyrin molecule, above and below) can result in irreversible conversion of Fe^{2+} to Fe^{3+}. In hemecontaining proteins, this reaction is prevented by sequestering of the heme deep within the protein structure where access to the two open coordination bonds is restricted. One of these two coordination bonds is occupied by a side-chain nitrogen of a His residue.

The other is the binding site for molecular oxygen (O_2). When oxygen binds, the electronic properties of heme iron change; this accounts for the change in color from the dark purple of oxygen-depleted venous blood to the bright red of oxygen-rich arterial blood.

Some small molecules, such as carbon monoxide (CO) and nitric oxide (NO), coordinate to heme iron with greater affinity than does O_2. When a molecule of CO is bound to heme, O_2 is excluded, which is why CO is highly toxic to aerobic organisms. By surrounding and sequestering heme, oxygenbinding proteins regulate the access of CO and other small molecules to the heme iron.

Myoglobin has a Single Binding Site for Oxygen

Myoglobin (M_r 16,700; abbreviated Mb) is a relatively simple oxygen-binding protein found in almost all mammals, primarily in muscle tissue. As a transport protein, it facilitates oxygen diffusion in muscle.

Myoglobin is particularly abundant in the muscles of diving mammals such as seals and whales, where it also has an oxygenstorage function for prolonged excursions undersea. Proteins very similar to myoglobin are widely distributed, occurring even in some single-celled organisms.

Myoglobin is a single polypeptide of 153 amino acid residues with one molecule of heme. It is typical of the family of proteins called globins, all of which have similar primary and tertiary structures. The polypeptide is made up of eight α-helical segments connected by bends. About 78% of the amino acid residues in the protein are found in these α helices.

Any detailed discussion of protein function inevitably involves protein structure. To facilitate our treatment of myoglobin, we first introduce some structural conventions peculiar to globins.

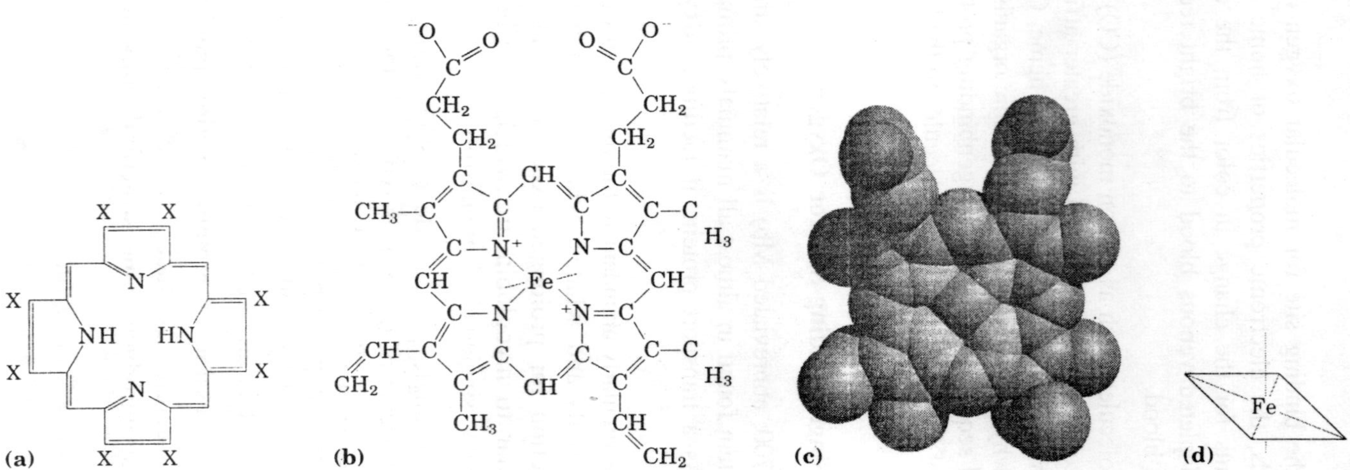

Figure. 26. Heme. (a) (a) Porphyrins, of which protoporphyrin IX is only one example, consist of four pyrrole rings linked by methene bridges, with substitutions at one or more of the positions denoted X. (b, c) Two representations of heme. The iron atom of heme has six coordination bonds: four in the plane of, and bonded to, the flat porphyrin ring system, and (d) two perpendicular to it.

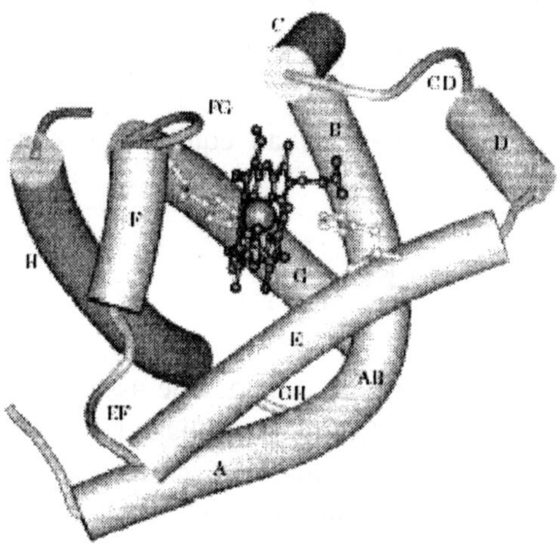

Figure.27. The structure of myoglobin.

An individual amino acid residue is designated either by its position in the amino acid sequence or by its location within the sequence of a particular α-helical segment. For example, the His residue coordinated to the heme in myoglobin, His^{93} (the 93rd amino acid residue from the amino-terminal end of the myoglobin polypeptide sequence), is also called His F8 (the 8th residue in α helix F). The bends in the structure are designated AB, CD, EF, FG, and so forth, reflecting the α-helical segments they connect.

Protein-Ligand Interactions

The function of myoglobin depends on the protein's ability not only to bind oxygen but also to release it when and where it is needed. Function in biochemistry often revolves around a reversible protein-ligand interaction of this type. A quantitative description of this interaction is therefore a central part of many biochemical investigations. In general, the reversible binding of a protein (P) to a ligand (L) can be described by a simple equilibrium expression:

$$P + L \rightleftharpoons PL \qquad (1)$$

The reaction is characterized by an equilibrium constant, K_a, such that

$$K_a \frac{[PL]}{[P][L]} \qquad (2)$$

The term K_a is an association constant. The association constant provides a measure of the affinity of the ligand L for the protein. K_a has units of M^{-1}; a higher value of K_a corresponds to a higher affinity of the ligand for the protein. A rearrangement of Equation 2 shows that the ratio of bound to free protein is directly proportional to the concentration of free ligand:

$$K_a [L] \frac{[PL]}{[P]} \qquad (3)$$

When the concentration of the ligand is much greater than the concentration of ligand-binding sites, the binding of the ligand by the protein does not appreciably change the concentration of free (unbound) ligand — that is, [L] remains constant. This condition is broadly applicable to most ligands that bind to proteins in cells and simplifies our description of the binding equilibrium. We can now consider the binding equilibrium from the standpoint of the fraction, θ (theta), of ligandbinding sites on the protein that are occupied by ligand:

$$\theta = \frac{\text{binding sites occupied}}{\text{total binding sites}} = \frac{[PL]}{[PL]+[P]} \qquad (4)$$

Substituting Ka[L][P] for [PL] and rearranging terms gives

$$\theta = \frac{K_a[L][P]}{K_a[L][P] + [P]} = \frac{K_a[L]}{K_a[L] + 1} = \frac{[L]}{[L] + \frac{1}{K_a}} \quad (5\text{–}5) \qquad (5)$$

The value of K_a can be determined from a plot of θ versus the concentration of free ligand, [L]. Any equation of the form $x = y/(y + z)$ describes a hyperbola, and θ is thus found to be a hyperbolic function of [L]. The fraction of ligand-binding sites occupied approaches saturation asymptotically as [L] increases. The [L] at which half of the available ligand-binding sites are occupied (at θ 0.5) corresponds to $1/K_a$.

It is more common (and intuitively simpler), however, to consider the dissociation constant, K_d, which is the reciprocal of K_a ($K_d = 1/K_a$) and is given in units of molar concentration (M). K_d is the equilibrium constant for the release of ligand. The relevant expressions change to

$$K_d = \frac{[P][L]}{[PL]} \qquad (6)$$

$$[PL] = \frac{[P][L]}{K_d} \qquad (7)$$

$$\theta = \frac{[L]}{[L] + K_d} \qquad (8)$$

When [L] is equal to K_d, half of the ligand-binding sites are occupied. As [L] falls below K_d, progressively less of the protein has ligand bound to it. In order for 90% of the available ligand-binding sites to be occupied, [L] must be nine times greater than K_d. In practice, K_d is used much more often than K_a to express the affinity of a protein for a ligand. Note that [L] a lower value of K_d corresponds to a higher affinity of ligand for the protein. The mathematics can be reduced to simple statements: K_d is equivalent to the molar concentration of ligand at which half of the available ligandbinding sites are occupied. At this point, the protein is said to have reached half-saturation with respect to ligand binding. The more tightly a protein binds a ligand, the lower the concentration of ligand required for half the binding sites to be occupied, and thus the lower the value of K_d. Some representative dissociation constants are given in Table 3.

Table. 3. Some Protein Dissociation Constants

Protein	*Ligand*	K_d *(M)**
Avidin (egg white)	Biotin	1×10^{-15}
Insulin receptor (human)	Insulin	1×10^{-10}
Anti-HIV immunoglobulin	gp41	4×10^{-10}
Nickel-binding protein	Ni^{2+}	1×10^{-7}
Calmodulin (rat)	Ca^{2+}	2×10^{-5}

The binding of oxygen to myoglobin follows the patterns discussed above. However, because oxygen is a gas, we must make some minor adjustments to the equations so that laboratory experiments can be carried out more conveniently. We first substitute the concentration of dissolved oxygen for [L] in Equation 5-8 to give

$$\theta = \frac{[O_2]}{[O_2] + K_d} \tag{9}$$

As for any ligand, K_d is equal to the $[O_2]$ at which half of the available ligand-binding sites are occupied, or $[O_2]_{0.5}$. Equation 9 thus becomes

$$\theta = \frac{[O_2]}{[O_2] + [O_2]_{0.5}} \tag{10}$$

In experiments using oxygen as a ligand, it is the partial pressure of oxygen in the gas phase above the solution, pO_2, that is varied, because this is easier to measure than the concentration of oxygen dissolved in the solution. The concentration of a volatile substance in solution is always proportional to the local partial pressure of the gas. So, if we define the partial pressure of oxygen at $[O_2]_{0.5}$ as P_{50}, substitution in Equation 10 gives

$$\theta = \frac{pO_2}{pO_2 + P_{50}} \qquad (11)$$

Protein Structure Affects how Ligands Bind

The binding of a ligand to a protein is rarely as simple as the above equations would suggest. The interaction is greatly affected by protein structure and is often accompanied by conformational changes. For example, the specificity with which heme binds its various ligands is altered when the heme is a component of myoglobin. Carbon monoxide binds to free heme molecules more than 20,000 times better than does O_2 (that is, the K_d or P_{50} for CO binding to free heme is more than 20,000 times lower than that for O_2), but it binds only about 200 times better when the heme is bound in myoglobin. The difference may be partly explained by steric hindrance.

When O_2 binds to free heme, the axis of the oxygen molecule is positioned at an angle to the FeOO bond. In contrast, when CO binds to free heme, the Fe, C, and O atoms lie in a straight line. In both cases, the binding reflects the geometry of hybrid orbitals in each ligand. In myoglobin, His^{64} (His E7), on the O_2-binding side of the heme, is too far away to coordinate with the heme iron, but it does interact with a ligand bound to heme. This residue, called the distal His, does not affect the binding of O_2 but may preclude the linear binding of CO, providing one explanation for the diminished binding of CO to heme in myoglobin (and hemoglobin). A reduction in CO binding is physiologically important, because CO is a lowlevel byproduct of cellular metabolism. Other factors, not yet well-defined, also seem to modulate the interaction of heme with CO in these proteins.

The binding of O_2 to the heme in myoglobin also depends on molecular motions, or "breathing," in the protein structure. The heme molecule is deeply buried in the folded polypeptide, with no direct path for oxygen to move from the surrounding solution to the ligandbinding site. If the protein were rigid, O_2 could not enter or leave the heme pocket at a measurable rate.

However, rapid molecular flexing of the amino acid side chains produces transient cavities in the protein structure, and O_2 evidently makes its way in and out by moving through these cavities. Computer simulations of rapid structural fluctuations in myoglobin suggest that there are many such pathways. One major route is provided by rotation of the side chain of the distal His (His^{64}), which occurs on a nanosecond (10^{-9} s) time scale Even subtle conformational changes can be critical for protein activity.

Oxygen is Transported in Blood by Hemoglobin

Oxygen-Binding Proteins-Hemoglobin: Oxygen Transport Nearly all the oxygen carried by whole blood in animals is bound and transported by hemoglobin in erythrocytes (red blood cells). Normal human erythrocytes are small (6 to 9 μ in diameter), biconcave disks. They are formed from precursor stem cells called hemocytoblasts.

In the maturation process, the stem cell produces daughter cells that form large amounts of hemoglobin and then lose their intracellular organelles-nucleus, mitochondria, and endoplasmic reticulum. Erythrocytes are thus incomplete, vestigial cells, unable to reproduce and, in humans, destined to survive for only about 120 days. Their main function is to carry hemoglobin, which is dissolved in the cytosol at a very high concentration (~34% by weight).

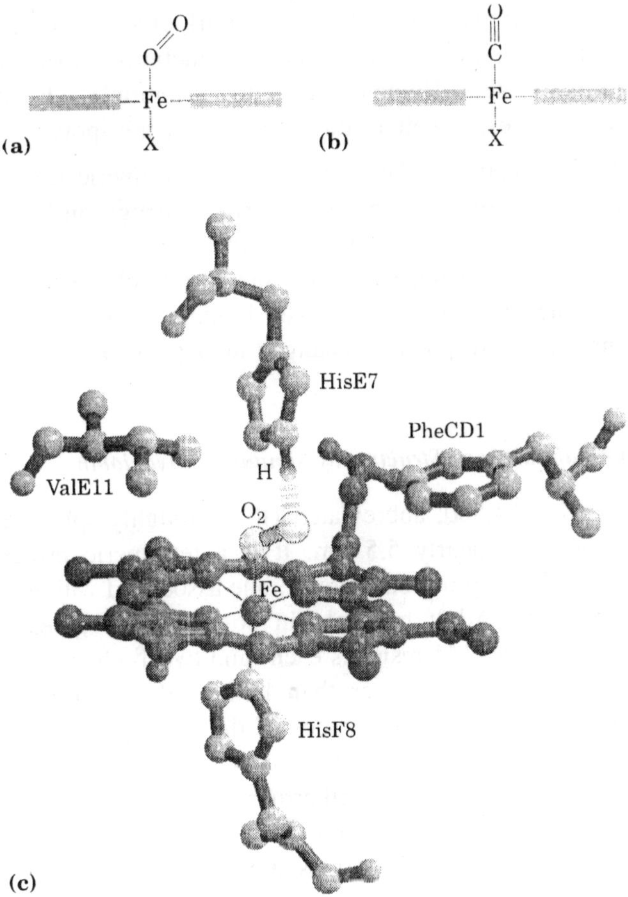

Figure.28. Steric effects on the binding of ligands to the heme of myoglobin

In arterial blood passing from the lungs through the heart to the peripheral tissues, hemoglobin is about 96% saturated with oxygen. In the venous blood returning to the heart, hemoglobin is only about 64% saturated. Thus, each 100 mL of blood passing through a tissue releases about one-third of the oxygen it carries, or 6.5 mL of O_2 gas at atmospheric pressure and body temperature. Myoglobin, with its hyperbolic binding curve for oxygen, is relatively insensitive to small changes in the concentration of dissolved oxygen and so functions well as an oxygen-storage protein. Hemoglobin, with its multiple subunits and O_2-binding sites, is better suited to oxygen transport.

interactions between the subunits of a multimeric protein can permit a highly sensitive response to small changes in ligand concentration. Interactions among the subunits in hemoglobin cause conformational changes that alter the affinity of the protein for oxygen. The modulation of oxygen binding allows the O_2-transport protein to respond to changes in oxygen demand by tissues.

Hemoglobin subunits are Structurally similar to Myoglobin

Hemoglobin (Mr 64,500; abbreviated Hb) is roughly spherical, with a diameter of nearly 5.5 nm. It is a tetrameric protein containing four heme prosthetic groups, one associated with each polypeptide chain. Adult hemoglobin contains two types of globin, two α chains (141 residues each) and two β chains (146 residues each). Although fewer than half of the amino acid residues in the polypeptide sequences of the α and α subunits are identical, the three-dimensional structures of the two types of subunits are very similar. Furthermore, their structures are very similar to that of myoglobin, even though the amino acid sequences of the three polypeptides are identical at only 27 positions.

All three polypeptides are members of the globin family of proteins. The helix-naming convention described for myoglobin is also applied to the hemoglobin polypeptides, except that the α subunit lacks the short D helix. The heme-binding pocket is made up largely of the E and F helices.

The quaternary structure of hemoglobin features strong interactions between unlike subunits. The $\alpha_1\beta_1$ interface (and its $\alpha_2\alpha_2$ counterpart) involves more than 30 residues, and its interaction is sufficiently strong that although mild treatment of hemoglobin with urea tends to cause the tetramer to disassemble into αβ dimers, these dimers remain intact. The α_1 β_2 (and α_2 β_1) interface involves 19 residues.

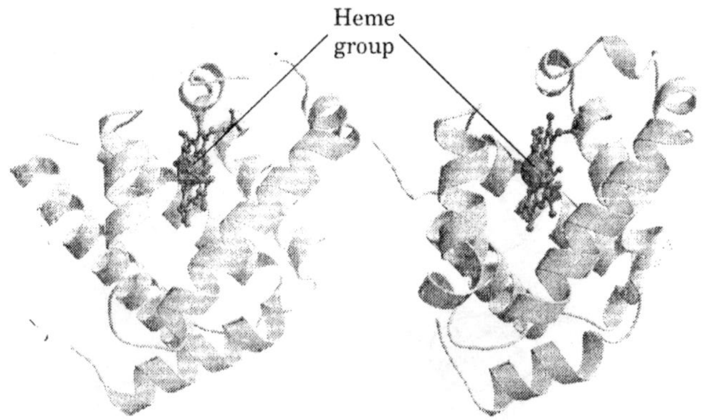

Figure.29. A comparison of the structures of myoglobin and the β subunit of the moglobin.

Hemoglobin undergoes a Structural Change on Binding Oxygen

X-ray analysis has revealed two major conformations of hemoglobin: the R state and the T state. Although oxygen binds to hemoglobin in either state, it has a significantly higher affinity for hemoglobin in the R state. Oxygen binding stabilizes the R state. When oxygen is absent experimentally, the T state is more stable and is thus the predominant conformation of deoxyhemoglobin. T and R originally denoted "tense" and "relaxed," respectively, because the T state is stabilized by a greater number of ion pairs, many of which lie at the $\alpha_1\beta_2$ (and $\alpha_2\beta_1$) interface. The binding of O_2 to a hemoglobin subunit in the T state triggers a change in conformation to the R state. When the entire protein undergoes this transition, the structures of the individual subunits change little, but the αβ subunit pairs slide past each other and rotate, narrowing the pocket between the β subunits. In this process, some of the ion pairs that stabilize the T state are broken and some new ones are formed.

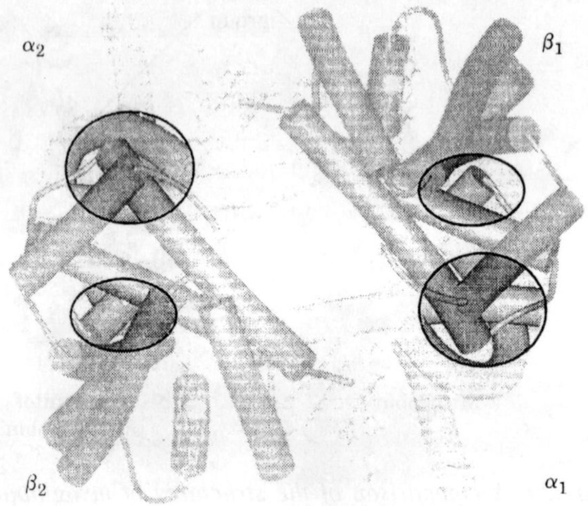

Figure.30. Dominant interactions between hemoglobin subunits

Max Perutz proposed that the T → R transition is triggered by changes in the positions of key amino acid side chains surrounding the heme. In the T state, the porphyrin is slightly puckered, causing the heme iron to protrude somewhat on the proximal His (His F8) side. The binding of O_2 causes the heme to assume a more planar conformation, shifting the position of the proximal His and the attached F helix. These changes lead to adjustments in the ion pairs at the $\alpha_1\alpha_2$ interface.

Hemoglobin Binds Oxygen Cooperatively

Hemoglobin must bind oxygen efficiently in the lungs, where the pO_2 is about 13.3 kPa, and release oxygen in the tissues, where the pO_2 is about 4 kPa. Myoglobin, or any protein that binds oxygen with a hyperbolic binding curve, would be ill-suited to this function. A protein that bound O_2 with high affinity would bind it efficiently in the lungs but would not release much of it in the tissues. If the protein bound oxygen with a sufficiently low affinity to release it in the tissues, it would not pick up much oxygen in the lungs.

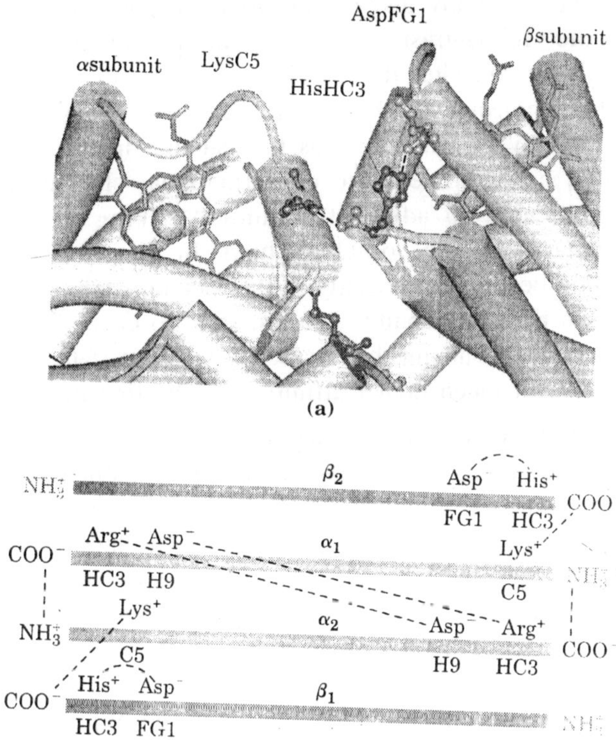

Figure.31. Some ion pairs that stabilisethe T state of deoxyhemoglobin.(a) A close-up view of a portion of a deoxyhemoglobin molecule in the T state. (b) The interactions between these ion pairs, and between others not shown in (a), are schematized in this representation of the extended polypeptide chains of hemoglobin.

Hemoglobin solves the problem by undergoing a transition from a low-affinity state (the T state) to a high-affinity state (the R state) as more O_2 molecules are bound. As a result, hemoglobin has a hybrid Sshaped, or sigmoid, binding curve for oxygen. A single-subunit protein with a single ligandbinding site cannot produce a sigmoid binding curve-even if binding elicits a conformational change-because each molecule of ligand binds independently and cannot affect the binding of another molecule. In contrast, O_2 binding to individual subunits of hemoglobin can alter the affinity for O_2 in adjacent subunits. The first molecule of O_2 that interacts with deoxyhemoglobin binds weakly, because it binds to a subunit in the T state. Its binding, however, leads to conformational changes that are communicated to adjacent subunits, making it easier for additional molecules of O_2 to bind. In effect, the T $\rightarrow$ R transition occurs more readily in the second subunit once O_2 is bound to the first subunit. The last (fourth) O_2 molecule binds to a heme in a subunit that is already in the R state, and hence it binds with much higher affinity than the first molecule.

An allosteric protein is one in which the binding of a ligand to one site affects the binding properties of another site on the same protein. The term "allosteric" derives from the Greek allos, "other," and stereos, "solid" or "shape." Allosteric proteins are those having "other shapes," or conformations, induced by the binding of ligands referred to as modulators. The conformational changes induced by the modulator(s) interconvert more-active and less-active forms of the protein. The modulators for allosteric proteins may be either inhibitors or activators. When the normal ligand and modulator are identical, the interaction is termed homotropic. When the modulator is a molecule other than the normal ligand the interaction is heterotropic. Some proteins have two or more modulators and therefore can have both homotropic and heterotropic interactions. Cooperative binding of a ligand to a multimeric protein, such as we observe with the binding of O_2 to hemoglobin, is a form of allosteric binding often observed in multimeric proteins.

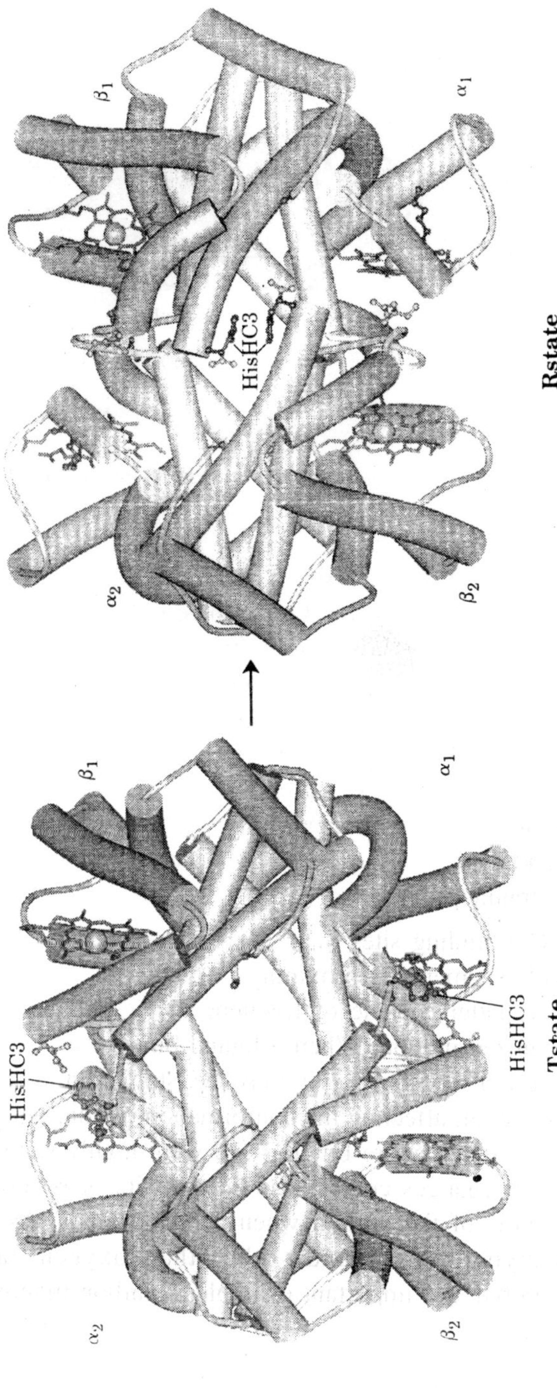

Figure.32. The T $\rightarrow$ R transition.

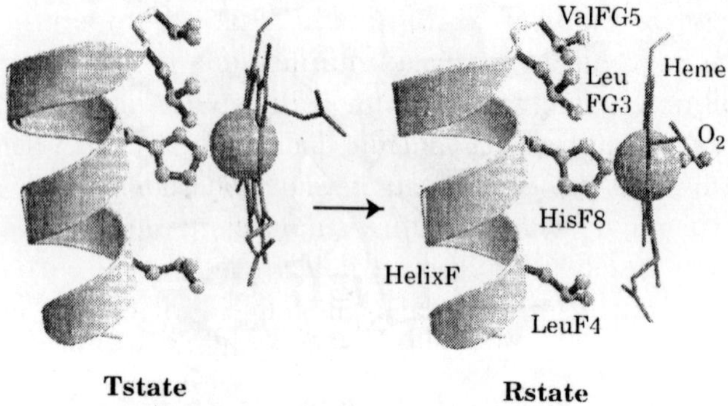

Figure.33. Changes in conformation near heme on O_2 binding to deoxyhemoglobin

The binding of one ligand affects the affinities of any remaining unfilled binding sites, and O_2 can be considered as both a ligand and an activating homotropic modulator. There is only one binding site for O_2 on each subunit, so the allosteric effects giving rise to cooperativity are mediated by conformational changes transmitted from one subunit to another by subunit-subunit interactions. A sigmoid binding curve is diagnostic of cooperative binding. It permits a much more sensitive response to ligand concentration and is important to the function of many multisubunit proteins.

The binding sites of an allosteric protein typically consist of stable segments in proximity to relatively unstable segments, with the latter capable of frequent changes in conformation or disorganized motion. When a ligand binds, the moving parts of the protein's binding site may be stabilized in a particular conformation, affecting the conformation of adjacent polypeptide subunits. If the entire binding site were highly stable, then few structural changes could occur in this site or be propagated to other parts of the protein when a ligand binds. As is the case with myoglobin, ligands other than oxygen can bind to hemoglobin. An important example is carbon monoxide, which binds to hemoglobin about 250 times better than does oxygen. Human exposure to CO can have tragic consequences.

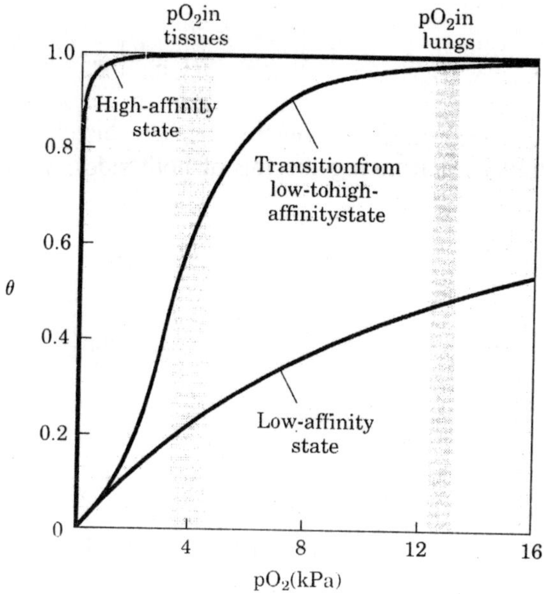

Figure.34. A sigmoid binding curve

Cooperative Ligand Binding

Cooperative binding of oxygen by hemoglobin was first analyzed by Archibald Hill in 1910. From this work came a general approach to the study of cooperative ligand binding to multisubunit proteins. For a protein with n binding sites, the equilibrium of Equation 1 becomes

$$P + nL \rightleftharpoons PL_n \tag{12}$$

and the expression for the association constant becomes

$$K_a = \frac{[PL_n]}{[P][L]^n} \tag{13}$$

The expression for θ is

$$\theta = \frac{[L]^n}{[L]^n + K_d} \tag{14}$$

Rearranging, then taking the log of both sides, yields

$$\frac{\theta}{1-\theta} = \frac{[L]^n}{K_d} \tag{15}$$

$$\log\left(\frac{\theta}{1-\theta}\right) = n \log [L] - \log K_d \tag{16}$$

where $K_d = [L]^u_{0.5}$.

Equation 16 is the Hill equation, and a plot of log θ / [(1– θ] versus log [L] is called a Hill plot. Based on the equation, the Hill plot should have a slope of *n*. However, the experimentally determined slope actually reflects not the number of binding sites but the degree of interaction between them. The slope of a Hill plot is therefore denoted by n_H, the Hill coefficient, which is a measure of the degree of cooperativity. If n_H equals 1, ligand binding is not cooperative, a situation that can arise even in a multisubunit protein if the subunits do not communicate. An n_H of greater than 1 indicates positive cooperativity in ligand binding. This is the situation observed in hemoglobin, in which the binding of one molecule of ligand facilitates the binding of others. The theoretical upper limit for n_H is reached when $n_H = n$. In this case the binding would be completely cooperative: all binding sites on the protein would bind ligand simultaneously, and no protein molecules partially saturated with ligand would be present under any conditions. This limit is never reached in practice, and the measured value of n_H is always less than the actual number of ligand-binding sites in the protein. An n_H of less than 1 indicates negative cooperativity, in which the binding of one molecule of ligand impedes the binding of others.

To adapt the Hill equation to the binding of oxygen to hemoglobin we must again substitute pO_2 for [L] and P_{50}^n for K_d:

$$\log\left(\frac{\theta}{1-\theta}\right) = n \log pO_2 - n \log P_{50}^n \qquad (5\text{–}17) \qquad (17)$$

Hill plots for myoglobin and haemoglobin are give in the following figure.

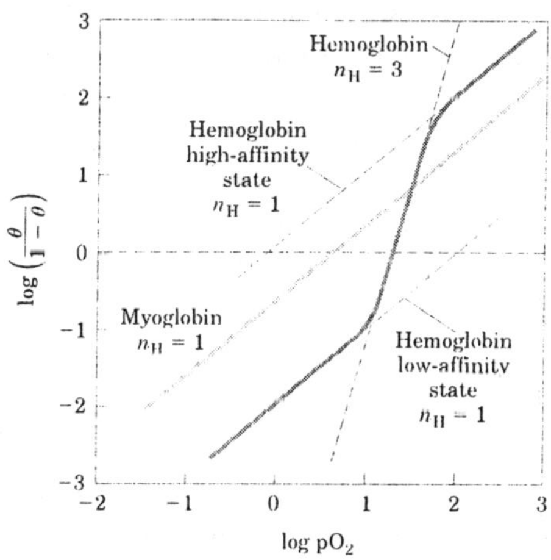

Figure.35. Hill plots for the binding of oxygen to myoglobin and hemoglobin. When n_H = 1, there is no evident cooperativity. The maximum degree of cooperativity observed for hemoglobin corresponds approximately to n_H = 3. While this indicates a high level of cooperativity, n_H is less than n, the number of O_2-binding sites in hemoglobin.

6

Synthesis of Protein

Protein biosynthesis is also called translation since it involves the biochemical translation of information from the four-letter language and structure of nucleic acids into the 20-letter language and structure of proteins. This process has many requirements: an informational messenger RNA molecule that is exported from the nucleus, several "bilingual" transfer RNA species that read the message, ribosomes that serve as catalytic and organizational centers, a variety of protein factors, and energy. Polypeptides are formed by the sequential addition of amino acids in the specific order determined by the information carried in the nucleotide sequence of the mRNA.

The protein is often then matured or processed by a variety of modifications. These may target it to a specific intracellular location or for secretion from the cell, or they may modulate its activity or function. These complex processes are carried out with considerable speed and extreme precision. Levels of translation are regulated, both globally and for specific proteins. Finally, when a protein becomes nonfunctional or is no longer needed, it is degraded and its amino acids are catabolized or recycled into new proteins.

Cells vary in their need and ability to synthesize proteins. At one extreme, terminally differentiated red blood cells have a life span of about 120 days, have no nuclei, do not divide, and do not synthesize proteins because they lack the components of the biosynthetic apparatus. Nondividing cells need to maintain levels of enzymes and other proteins and carry out limited protein synthesis. Growing and dividing cells must synthesize much larger amounts of protein. Finally, some cells synthesize proteins for export as well as for their own use. For example, liver cells synthesize large numbers of enzymes needed for their many metabolic pathways as well as proteins for export, including serum albumin, the major protein of blood plasma or serum. Liver cells are protein factories that are particularly rich in the machinery for synthesis of proteins.

Components of the Transational Apparatus

Genetic information is stored and transmitted in the nucleotide sequences of DNA. Selective expression of this information requires its transcription into mRNA that carries specific and precise messages from the nuclear "data bank" to the cytoplasmic sites of protein synthesis. In eukaryotes, the messengers, mRNAs, are usually synthesized as significantly larger precursor molecules that are processed prior to export from the nucleus. Eukaryotic mRNA in the cytosol has several identifying characteristics. It is almost always monocistronic, that is, encoding a single polypeptide. The 59 end is capped with a specific structure consisting of 7-methylguanosine linked through a 59-triphosphate bridge to the 59 end of the messenger sequence.

A 59-nontranslated region, which may be short or up to a few hundred nucleotides in length, separates the cap from the translational initiation signal, an AUG codon. Usually, but not always, this is the first AUG sequence encountered as the message is read 59 R 39. Uninterrupted sequences that specify a unique polypeptide sequence follow the initiation signal until a specific translation termination signal is reached. This is followed by a 39-untranslated sequence, usually about 100 nucleotides in length, before the mRNA is terminated by a 100- to 200-nucleotide long polyadenylate tail. Prokaryotic mRNA differs from eukaryotic mRNA in that the 59 terminus is not capped but retains a terminal triphosphate from initiation of its synthesis by RNA polymerase. Also, most messengers are polycistronic, that is, encoding several polypeptides, and include more than one initiation AUG sequence. A ribosome-positioning sequence is located about 10 nucleotides upstream of a valid AUG initiation signal. An untranslated sequence follows the termination signal, but there is no polyadenylate tail.

Ribosomes are workbenches for protein biosynthesis

Proteins are assembled on particles called ribosomes. These have two dissimilar subunits, each of which contains RNA and many proteins. With one exception, each protein is present in a single copy per ribosome, as is each RNA species. The composition of major ribosome types is shown in Table 1.

Ribosome architecture has been conserved in evolution. The similarities between ribosomes and subunits from different sources are more obvious than the differences, and functional roles for each subunit are well defined. Details of ribosome structure and its relationship to function have been learned using many techniques. Overall size and shape can be determined by electron microscopy. The location of many ribosomal proteins, some elements of the RNA, and functional sites on each subunit have been determined by electron microscopy of subunits that are complexed with antibodies against a single ribosomal component. The antibody molecule serves as a physical pointer to the site on the ribosome. Further structural information has been obtained from chemical cross-linking, which identifies near neighbors within the structure, and from neutron diffraction measurements, which quantitate the distances between pairs of proteins. Ribosomes have been crystallized and X-ray structural determination is under way. Sequence comparisons and chemical, immunological, and enzymatic probes give information about RNA conformation. Correlations of structural data with functional measurements in protein synthesis have allowed development of models, such as that in Figure 1, that link ribosome morphology to various functions in translation. Each subunit has an RNA core, folded into a specific three-dimensional structure, upon which proteins are positioned through protein-RNA and protein-protein interactions.

Table 1. Ribosome Classification and Composition

Ribosome Source	*Monomer Size*	*Subunits*	
		Small	*Large*
Eukaryotes			
Cytosol	80S	40S: 34 proteins 18S RNA	60S: 50 proteins 28S, 5.8S, 5S RNAs
Mitochondria			
Animals	55S–60S	30S–35S: 12S RNA	40–45S: 16S RNA 70–100 proteins
Higher plants	77S–80S	40S: 19S RNA	60S: 25S, 5S RNAs 70–75 proteins
Chloroplasts	70S	30S: 20–24 proteins 16S RNA	50S: 34–38 proteins 23S, 5S, 4.5S RNAs
Prokaryotes			
Escherichia coli	70S	30S: 21 proteins 16S RNA	50S: 34 proteins 23S, 5S RNAs

Many of these experiments were possible because prokaryotic ribosomes can self-assemble; that is, the native structures can be reconstituted from mixtures of purified individual proteins and RNAs. Reconstitution of subunits from mixtures in which a single component is omitted or modified can show, for example,

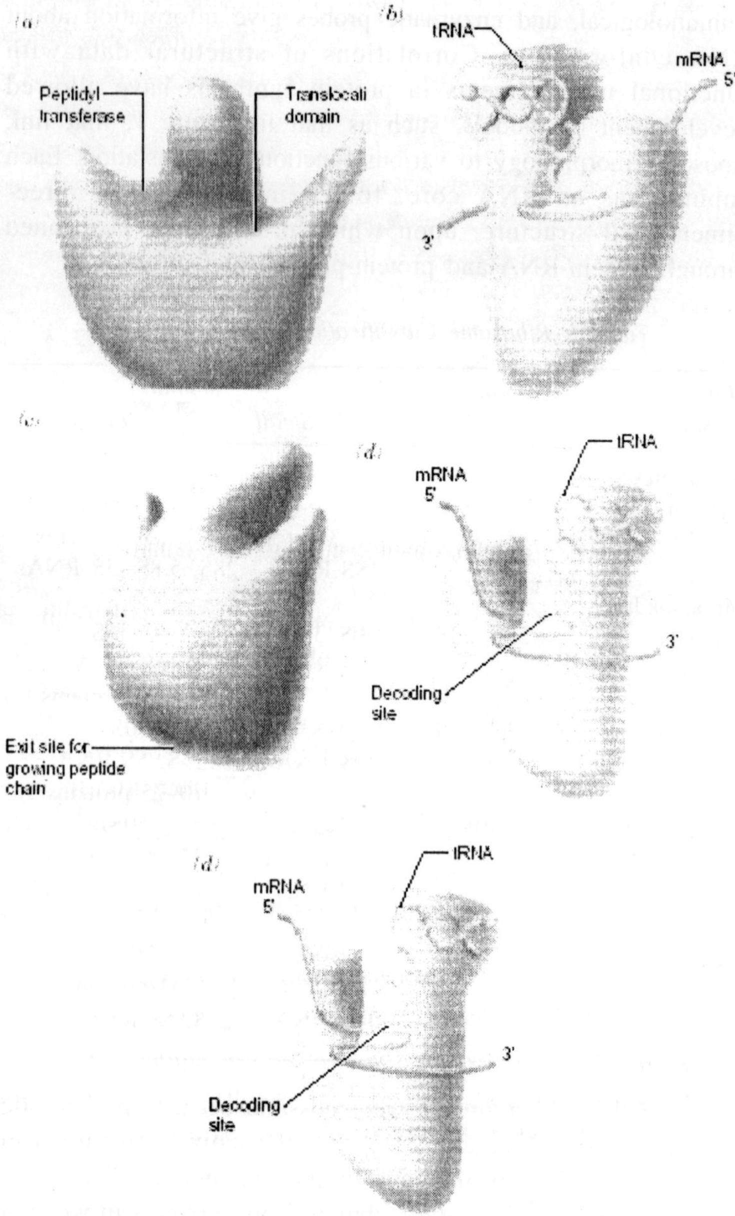

Figure.1. Ribosome structure and functional sites.

if a given protein is required for assembly of the subunit or for some specific function. An assembly map for large ribosomal subunits of Escherichia coli is shown in Figure 2. Total reconstitution of subunits from eukaryotes has not yet been achieved but the general conclusions about how ribosomes function, although determined using bacterial ribosomes, are fully applicable to eukaryotic systems. Ribosomes are organized in two additional ways.

First, several ribosomes often translate a single mRNA molecule simultaneously. Purified mRNA-linked polysomes can be visualized by electron microscopy. Second, in eukaryotic cells some ribosomes occur free in the cytosol, but many are bound to membranes of the rough endoplasmic reticulum. In general, free ribosomes synthesize proteins that remain within the cell cytosol or become targeted to the nucleus, mitochondria, or some of the other organelles. Membrane-bound ribosomes synthesize proteins that will be secreted from the cell or sequestered and function in other cellular membranes or vesicles. In cell homogenates, membrane fragments and the bound ribosomes constitute the microsome fraction; detergents that disrupt membranes release these ribosomes.

Transfer RNA Acts as a Bilingual Translator Molecule

All tRNA molecules have several common structural characteristics including the 39-terminal CCA sequence to which amino acids are bound, a highly conserved cloverleaf secondary structure, and an L-shaped three-dimensional structure (see p. XXX). But each of the many molecular species has a unique nucleotide sequence, giving it individual characteristics that allow great specificity in inter-actions with mRNA and with the aminoacyl-tRNA synthetase that couples one specific amino acid to it.

The Genetic Code Uses a Four-Letter Alphabet of Nucleotides

Information in the cell is stored in the form of linear sequences of nucleotides in DNA, in a manner that is analogous to the linear sequence of letters of the alphabet in the words you are now reading. The DNA language uses a simple four-letter alphabet that comprises the two purines, A and G (adenine and guanine), and the two pyrimidines, C and T (cytosine and thymine). In mRNA the information is encoded in a similar four-letter alphabet, but U (uracil) replaces T. The language of RNA is thus a dialect of the genetic language of DNA. Genetic information is expressed predominantly in the form of proteins that derive their properties from their linear sequence of amino acids and to a much lesser extent as RNA species such as tRNA and rRNA. Thus, during protein biosynthesis, the four-letter language of nucleic acids is translated into the 20-letter language of proteins. Implicit in the analogy to language is the directionality of these sequences. By convention, nucleic acid sequences are written in a 5' → 3' direction, and protein sequences from the amino terminus to the carboxy terminus. These directions inmRNAand protein correspond in both their reading and biosynthetic senses.

Codons in mRNA Are Three-Letter Words

A 1 : 1 correspondence of nucleotides to amino acids would only permit mRNA to encode four amino acids, while a 2 : 1 correspondence would encode $4^2 = 16$ amino acids. Neither is sufficient since 20 amino acids occur in most proteins. The actual three-letter genetic code has $4^3 = 64$ permutations or words, which is also sufficient to encode start and stop signals, equivalent to punctuation.

Only two amino acids are designated by single codons: methionine as AUG and tryptophan as UGG. The rest are designated by two, three, four, or six codons. Multiple codons for a single amino acid represent degeneracy in the code. The genetic code is nearly universal. The same code words are used in all living organisms, prokaryotic and eukaryotic. An exception to universality occurs in mitochondria, in which a few codons have a different meaning than in the cytosol of the same organism.

Table 2. Nonuniversal Codon Usage in Mammalian Mitochondria

Codon	*Usual Code*	*Mitochondrial Code*
UGA	Termination	Tryptophan
AUA	Isoleucine	Methionine
AGA	Arginine	Termination
AGG	Arginine	Termination

Punctuation

Four codons function partly or totally as punctuation, signaling the start and stop of protein synthesis. The start signal, AUG, also specifies methionine. An AUG at an appropriate site and within an acceptable sequence in mRNA signifies methionine as the initial, amino-terminal residue. AUG codons elsewhere in the message specify methionine residues within the protein. Three codons, UAG, UAA, and UGA, are stop signals; they specify no amino acid and are known as termination codons or, less appropriately, as nonsense codons.

Codon-Anticodon Interactions Permit Reading of mRNA

Translation of the codons of mRNA involves their direct interaction with complementary anticodon sequences in tRNA. Each tRNA species carries a unique amino acid, and each has a specific three-base anticodon sequence. Codon-anticodon base pairing is antiparallel, as shown in Figure 2. The anticodon is far from the amino acid-acceptor stem in both the tRNA cloverleaf and the L-shaped three-dimensional structure of all tRNA molecules. Location of the anticodon and amino acid residue at opposite extremes of the molecule permits the tRNA to conceptually and physically bridge the gap between the nucleotide sequence of the ribosome-bound mRNA and the site of protein assembly on the ribosome.

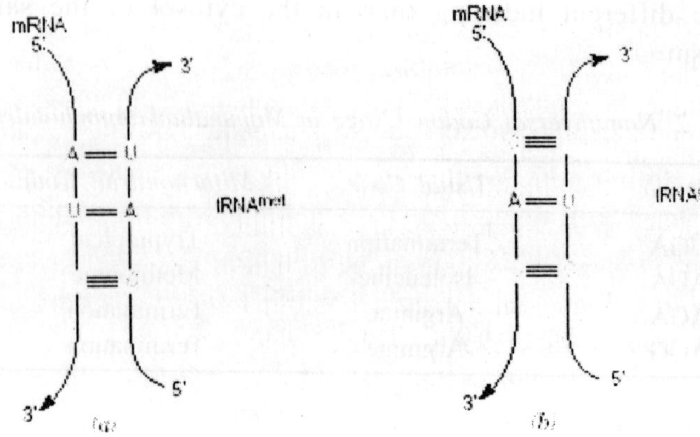

Figure.2. Codon–anticodon interactions.

Since 61 codons designate an amino acid, it might seem necessary to have 61 different tRNA species. This is not the case. Variances from standard base pairing are common in codon-anticodon interactions. Many amino acids can be carried by more than one tRNA species, and degenerate codons can be read by more than one tRNA (but always one carrying the correct amino acid). Much of this complexity is explained by the "wobble" hypothesis, which permits less stringent base pairing between the third position of a codon and the first position of its anticodon. Thus the first two positions of a codon predominate in tRNA selection and the degenerate (third) position is less important. A second modulator of codon-anticodon interactions is the presence of modified nucleotides at or beside the first nucleotide of the anticodon in many tRNA species. A frequent anticodon nucleotide is inosinic acid (I), the nucleotide of hypoxanthine, which base pairs with U, C, or A. Wobble base pairing rules are shown in Table 3.

If the wobble rules are followed, the 61 nonpunctuation codons could be read by as few as 31 tRNA molecules, but most cells have 50 or more tRNA species. Some codons are read more efficiently by one anticodon than another. Not all codons are used equally, some being used very rarely. Examination of many mRNA sequences has allowed construction of "codon usage" tables that show that different organisms preferentially use different codons to generate similar polypeptide sequences.

Table 3. Wobble Base Pairing Rules

3' Codon Base	*5' Anticodon Bases Possible*
A	U or I
C	G or I
G	C or U
U	A or G or I

"Breaking" the Genetic Code

The genetic code was determined before methods were developed to sequence natural mRNA. These code-breaking experiments provide insight into how proteins are synthesized. Important experiments used simple artificial mRNAs or chemically synthesized trinucleotide codons.

Polynucleotide phosphorylase catalyzes the template-independent and readily reversible reaction:

$$x\text{NDP} \leftrightarrows \text{polynucleotide } (\text{pN})_x + \text{P}_i$$

where NDP is any nucleoside 59-diphosphate or a mixture of two or more. If the nucleoside diphosphate is UDP, a polymer of U designated poly(U) is formed. Under nonphysiological conditions protein synthesis can occur in vitro without the initiation components that are normally required. With poly(U) as mRNA, the "protein" polyphenylalanine is made. Similarly, poly(A) encodes polylysine and poly(C) polyproline. An mRNA with a random sequence of only U and C produces polypeptides that contain not only proline and phenylalanine, as predicted, but also serine (from UCU and UCC) and leucine (from CUU and CUC). Because of degeneracy in the code and the complexity of the products, experiments with random sequence mRNAs were difficult to interpret, and so synthetic messengers of defined sequence were transcribed from simple repeating DNA sequences by RNA polymerase.

Thus poly(AU), transcribed from a repeating poly(dAT), produces only a repeating copolymer of Ile-Tyr-Ile-Tyr, read from successive triplets AUA UAU AUA UAU and so on. A synthetic poly(CUG) has possible codons CUG for Leu, UGC for Cys, and GCU for Ala, each repeating itself once the reading frame has been selected. Since selection of the initiation codon is random in these in vitro experiments, three different homopolypeptides are produced: polyleucine, polycysteine, and polyalanine.

Other experiments used chemically synthesized trinucleotide codons as minimal messages. No proteins were made, but the binding of only one amino acid (conjugated to an appropriate tRNA) by the ribosome was stimulated by a given codon. It was thus possible to decipher the meaning of each possible codon and to identify termination codons. All of these conclusions were later verified by the determination of mRNA sequences.

Mutations

An understanding of the genetic code and how it is read provides a basis for understanding the nature of mutations. A mutation is simply a change in a gene. Point mutations involve a change in a single base pair in the DNA, and thus a single base in the corresponding mRNA. Sometimes this change occurs in the third position of a degenerate codon and there is no change in the amino acid specified (e.g., UCC to UCA still codes for serine). Such silent mutations are only detected by gene sequence determination. They are commonly seen during comparison of genes for similar proteins, for example, hemoglobins from different species. Missense mutations arise from a base change that causes incorporation of a different amino acid in the encoded protein. Point mutations can also form or destroy a termination codon and thus change the length of a protein. Formation of a termination codon from one that encodes an amino acid is often called a nonsense mutation; it results in premature termination and a truncated protein. Mutation of a termination codon to one for an amino acid allows the message to be "read through" until another stop codon is encountered. The result is a larger than normal protein. This phenomenon is the basis of several disorders. Insertion or deletion of a single nucleotide within the coding region of a gene results in a frameshift mutation. The reading frame is altered at that point and subsequent codons are read in the new context until a termination codon is reached. The significance of reading frame selection is underscored by a phenomenon in some viruses in which a single segment of DNA encodes different polypeptides that are translated using different reading frames. An example is the tumor-causing simian virus SV40, whose small size physically limits the amount of DNA that can be packaged within it.

Aminoacylation of Transfer RNA Activates

In order to be incorporated into proteins, amino acids must first be "activated" by linkage to their appropriate tRNA carriers. This is a two-step process that requires energy and is catalyzed by one of a family of aminoacyl-tRNA synthetases, each of which is specific for a single amino acid and its appropriate tRNA species.

The brackets surrounding the aminoacyl-AMP-enzyme complex indicate that it is a transient, enzyme-bound intermediate. The "squiggle" (~) linkage of amino acid to AMP identifies the aminoacyl-adenylate as a high-energy intermediate, a mixed acid anhydride with carboxyl and phosphoryl components. The aminoacyl ester linkage in tRNA is lower in energy than the aminoacyl-adenylate, but still higher than that of the carboxyl group of the free amino acid. The reactions are written to show their reversibility. In reality, pyrophosphatases cleave the pyrophosphate released and the equilibrium is strongly shifted toward formation of aminoacyl-tRNA. From the viewpoint of precision in translation, the amino acid, which had only its side chain (R group) to distinguish it, becomes linked to a large, complex, and easily recognized carrier.

Specificity and Fidelity of Aminoacylation Reactions

Cells contain 20 different aminoacyl-tRNA synthetases, each specific for one amino acid, and at most a small family of carrier tRNAs for that amino acid. In translation, codon-anticodon interactions define the amino acid to be incorporated. If an incorrect amino acid is carried by the tRNA, it will be incorporated into the protein. Correct selection of both tRNA and amino acid by the synthetase is necessary to avoid such mistakes. Accuracy of these enzymes is central to the fidelity of protein synthesis.

Table 4

"Read Through" Mutation in Termination Codons Produce Abnormally Long a-Globin Chains

Hemoglobin	*α-Codon 142*	*Amino Acid 142*	*α-Globin Length (Residues)*
A		UAA	141
Constant Spring	CAA	Glutamine	172
Icaria	AAA	Lysine	172
Seal Rock	GAA	Glutamate	172
Koya Dora	UCA	Serine	172

Aminoacyl-tRNA synthetases share a common mechanism and many are physically associated with one another in the cell. Nevertheless, they are a diverse group of proteins that may contain one, two, or four identical subunits or pairs of dissimilar subunits. Detailed studies indicate that separate structural domains are involved in aminoacyl-adenylate formation, tRNA recognition, and, if it occurs, subunit interactions. In spite of their structural diversity, each enzyme is capable of almost error-free formation of correct aminoacyl-tRNA combinations.

Selection and incorporation of a correct amino acid require great discrimination on the part of some synthetases. While some amino acids be may easily recognized by their bulk (e.g., tryptophan) or lack of bulk (glycine), or by positive or negative charges on the side chains (e.g., lysine and glutamate), others are much more difficult to discriminate. Recognition of valine rather than threonine or isoleucine by the valyl-tRNA synthetase is difficult since the side chains differ by either an added hydroxyl or single methylene group. The amino acid-recognition and -activation sites of each enzyme have great specificity, as is characteristic of many enzymes. Nevertheless, misrecognition does occur.

This most often occurs through hydrolysis of the aminoacyl-adenylate intermediate, with the release of amino acid and AMP. Valyl-tRNA synthetase efficiently hydrolyzes threonyl-adenylate and it hydrolyzes isoleucyl-adenylate in the presence of bound (but not aminoacylated) $tRNA^{Val}$. In other cases a misacylated tRNA is recognized and deacylated. Valyl- and phenylalanyl-tRNA synthetases deacylate tRNAs that have been mischarged with threonine and tyrosine, respectively. This proofreading is analogous to editing of misincorporated nucleotides by the 3'→ 5' exonuclease activity of DNA polymerases. Editing is performed by many but not all aminoacyl-tRNA synthetases. The net result is an average level of misacylation of one in 10^4 to 10^5.

Each synthetase must correctly recognize one to several tRNA species that correctly serve to carry the same amino acid, while rejecting incorrect tRNA species. Given the complexity of tRNA molecules, this should be simpler than selection of a single amino acid. However, recall the conformational similarity and common sequence elements of all tRNAs. Different synthetases recognize different elements of tRNA structure. One logical element of tRNA recognition by the synthetase is the anticodon, specific to one amino acid. For example, in the case of $tRNA^{Met}$, changing the anticodon also alters recognition by the synthetase. In other instances, this is at least partly true. Sometimes the anticodon is not a determinant of synthetase-tRNA recognition. Consider, for example, suppressor mutations that "suppress" the expression of classes of chain termination (nonsense) mutations. A point mutation in a glutamine (CAG) codon produces a termination (UAG) codon, which causes the premature termination of the encoded protein.

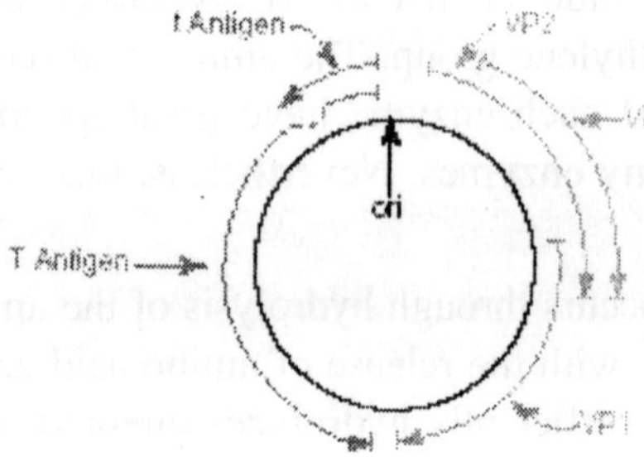

Figure.3. Map of genome of simian virus 40

A second suppressor mutation in the anticodon of a tRNATyr, in which the normal GUA anticodon is changed to CUA, allows "read through" of the termination codon. The initial mutation is suppressed as a nearly normal protein is made, with the affected glutamine replaced by tyrosine. Aminoacylation of the mutant $tRNA^{tyr}$ with tyrosine shows that in this case the anticodon does not determine synthetase specificity. In E. coli $tRNA^{Ala}$, the primary recognition characteristic is a G_3-U_{70} base pair in the acceptor stem; even if no other changes in the $tRNA^{Ala}$ occur, any variation at this position destroys its acceptor ability with alanine-$tRNA^{Ala}$ synthetase. Incorporation of a G_3-U_{70} base pair in tRNACys makes it an alanine acceptor, and even the isolated acceptor stem of $tRNA^{Ala}$ can be aminoacylated.

Other tRNA identification features include additional elements of the acceptor stem and sometimes parts of the variable loop or the D-stem/loop. Usually multiple structural elements contribute to recognition, but many are not absolute determinants. The X-ray structure of the glutaminyl synthetase-tRNA complex shown in Figure 4 shows binding at the concave tRNA surface, which is typical and compatible with the biochemical observations.

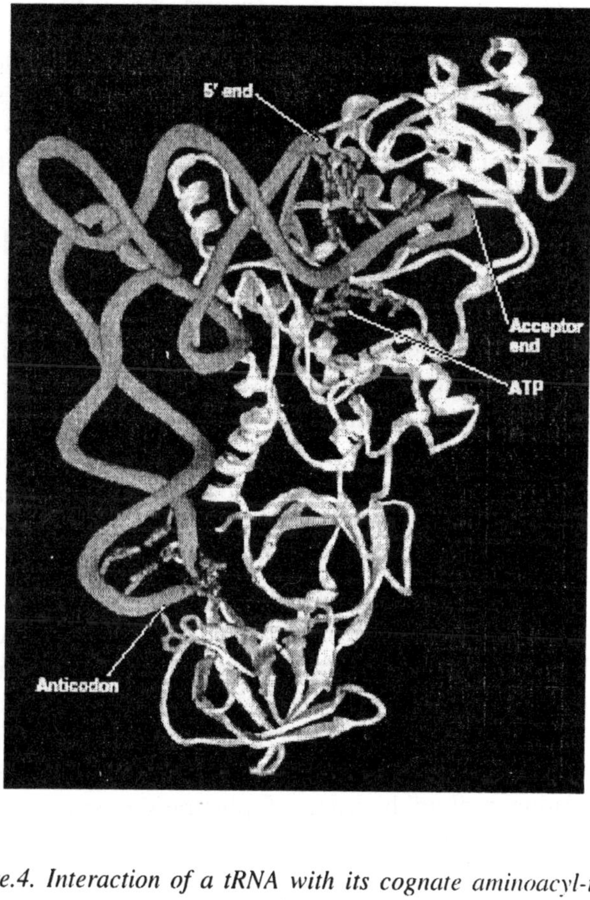

Figure.4. Interaction of a tRNA with its cognate aminoacyl-tRNA synthetase

Protein Biosynthesis

Translation Is Directional and Colinear with mRNA

In the English language words are read from left to right and not from right to left. Similarly, mRNA sequences are written 5' → 3' and in the translation process they are read in the same direction. Amino acid sequences are both written and biosynthesized from the amino-terminal residue to the carboxy terminus. This was first demonstrated by following the incorporation of radioactive amino acids into specific sites in hemoglobin as a function of time. Only full length, complete globin chains were isolated and analyzed. Completed chains that incorporated radioactive amino acids during the shortest exposures to the radioactive precursor were near to being finished at the time of the pulse and were found to have radioactive amino acids only in the carboxy-terminal segments.

Longer pulses with radioactive amino acids resulted also in labeling of central segments of the protein, and the longest pulse time, still corresponding to less than that needed to synthesize a full-length polypeptide, showed radioactivity approaching the amino-terminal segments. Again, this amino- to carboxy-terminal directionality became obvious as details of translation were clarified. The existence of stable polysomes and the directional nature of translation imply that each ribosome remains bound to an mRNA molecule and moves along the length of the mRNA until it is fully read. Comparison of mRNA sequences with sequences of the proteins they encode shows a perfect, colinear, gap-free correspondence of the mRNA coding sequence and that of the synthesized polypeptide. In fact, it is common to deduce the sequence of a protein solely from the nucleotide sequence of its mRNA or the DNA of the gene encoding it. However, the deduced sequence may differ from the genuine protein because of posttranslational events and modifications.

Initiation of Protein Synthesis Is a Complex Process

A good novel can be analyzed in terms of its beginning, its development or middle section, and its satisfactory ending. Protein biosynthesis will be described in a similar conceptual and mechanical framework: initiation of the process, elongation during which the great bulk of the protein is formed, and termination of synthesis and release of the finished polypeptide. We will then examine the posttranslational modifications that a protein may undergo.

Initiation requires bringing together a small (40S) ribosomal subunit, the mRNA, and a tRNA complex of the amino-terminal amino acid, all in a proper orientation. This is followed by association of the large (60S) subunit to form a completed initiation complex on an 80S ribosome. The ordered process is shown in Figure 17.7; it also requires a complex group of proteins, known as initiation factors, that participate only in initiation. They are not ribosomal proteins, although many of them bind transiently to ribosomes during initiation steps. There are many eukaryotic initiation factors and the specific functions of some remain unclear; prokaryotic protein synthesis provides a useful and less complex model for comparison.

As a first step, eukaryotic initiation factor 2a (eIF-2a) binds to GTP and one species of $tRNA^{Met}_{i}$, designated $Met\text{-}tRNA^{Met}_{i}$, to form a ternary complex. Recognition of the initiator tRNA by eIF-2a is in some ways analogous to the recognition of a tRNA species by its cognate aminoacyltransferase. No other aminoacyl-tRNA, including $Met\text{-}tRNA^{Met}_{c}$, which participates only in elongation steps of protein synthesis, normally replaces the initiation-specific $Met\text{-}tRNA^{Met}_{i}$ in this step.

Prokaryotes also utilize a specific initiator tRNA whose methionine is modified by formylation of its amino group. Only $fMet\text{-}tRNA^{Met}_{i}$ is recognized by prokaryotic IF-2. The second step in initiation requires 40S ribosomal subunits associated with a very complex protein, eIF-3. Mammalian eIF-3 includes eight different polypeptides and has a mass of 600-650 kDa. In electron micrographs eIF-3 is seen bound to the 40S subunit surface that will contact the larger 60S subunit, thus physically blocking association of 40S and 60S subunits. Hence eIF-3 is also called a ribosome anti-association factor, as is eIF-6, which binds to 60S subunits. A complex that includes eIF-2a ? Met-$tRNA^{Met}_i$ · GTP, eIF-3 · 40S, eIF-4c, and perhaps additional protein factors now forms. This complex is able to bind mRNA and thus generate a preinitiation complex, but once again several protein factors play a role.

Proteins eIF-4a and eIF-4b serve to unwind or melt secondary structure in the mRNA, and eIF-4f (or cap binding protein) helps place the message on the 40S subunit with the capped 59 end correctly oriented. Hydrolysis of ATP is required in "scanning" up to 100 or more nucleotides of the mRNA in order to locate and place the initiation AUG sequence in the correct site on the 40S subunit. Usually the first AUG sequence in the mRNA is used, but sometimes the surrounding nucleotide sequence or secondary structure is not appropriate for initiation and a later AUG is selected. The final preinitiation complex thus includes the 40S subunit, the eIF-2a · Met-$tRNA^{Met}_i$ · GTP ternary complex, correctly oriented mRNA, and several protein factors.

Formation of the complete initiation complex now proceeds with involvement of a 60S subunit and an additional factor, eIF-5. Protein eIF-5 first interacts with the preinitiation complex; GTP is hydrolyzed to GDP and Pi , and eIF-2a · GDP, eIF-3, and other factors are released. The 40S · Met-$tRNA^{Met}$ · mRNA complex interacts with a 60S subunit and initiation factor eIF-4d to generate an 80S ribosome with the mRNA and initiator tRNA correctly positioned on the ribosome. The eIF-2a · GDP that is released interacts with the guanine nucleotide exchange factor eIF-2b and GTP to regenerate eIF-2a · GTP for another round of initiation.

Prokaryotes use fewer nonribosomal factors and a slightly different order of interaction. Their 30S subunits complexed with a simpler IF-3 first bind mRNA. Orientation of the mRNA relies in part on base pairing between a pyrimidine-rich sequence of eight nucleotides in 16S rRNA and a purine-rich

"Shine-Dalgarno" sequence (named for its discoverers) about 10 nucleotides upstream of the initiator AUG codon. Complementarity between rRNA and the message-positioning sequence of an mRNA may include several mismatches but, as a first approximation, the better the complementary pairing the more efficient initiation at that AUG will be. It is interesting that eukaryotes do not utilize an mRNA-rRNA base pairing mechanism, but instead use many protein factors to position mRNA correctly. After the mRNA is bound by a 30S subunit, a ternary complex of IF-2, fMet-$tRNA^{Met}_{i}$, and GTP is bound. A third initiation factor, IF-1, also participates in formation of the preinitiation complex. A 50S subunit is now bound; in the process, GTP is hydrolyzed to GDP and Pi , and the initiation factors are released.

Elongation Is the Stepwise Formation of Peptide Bonds

Protein synthesis now occurs by stepwise elongation to form a polypeptide chain. At each step ribosomal peptidyltransferase transfers the growing peptide (or in the first step the initiating methionine residue) from its carrier tRNA to the a-amino group of the amino acid residue of the aminoacyl-tRNA specified by the next codon. Efficiency and fidelity are enhanced by nonribosomal protein elongation factors that utilize the energy released by GTP hydrolysis to ensure selection of the proper aminoacyl-tRNA species and to move the mRNA and associated tRNAs through the decoding region of the ribosome.

At a given moment, up to three different tRNA molecules may be bound at specific sites that span both ribosomal subunits. The initiating methionyltRNA is placed in position so that its methionyl residue may be transferred (or donated) to the free a-amino group of the incoming aminoacyl-tRNA; it thus occupies the donor site, also called the peptidyl site or P site of the ribosome. The aminoacyl-tRNA specified by the next codon of the message is bound at the acceptor site, also called the aminoacyl site or A site of the ribosome. Selection of the correct aminoacyl-tRNA is enhanced by elongation factor 1 (EF-1); a component of EF-1, EF-1α, first forms a ternary complex with aminoacyl-tRNA and GTP. The EF-1α · aminoacyl-tRNA · GTP complex binds to the ribosome and if codon-anticodon interactions are correct, the aminoacyl-tRNA is placed at the A site, GTP is hydrolyzed to GDP and Pi, and the EF-1α · GDP complex dissociates.

The initiating methionyl-tRNA and the incoming aminoacyl-tRNA are now juxtaposed on the ribosome. Their anticodons are paired with successive codons of the mRNA in the decoding region of the small subunit, and their amino acids are beside one another at the peptidyltransferase site of the large subunit. Peptide bond formation now occurs. Peptidyltransferase catalyzes the attack of the a-amino group of the aminoacyl-tRNA onto the carbonyl carbon of the methionyl-tRNA. The result is transfer of the methionine to the amino group of the aminoacyl-tRNA, which then occupies a "hybrid" position on the ribosome. The anticodon remains in the 40S A site, while the acceptor end and the attached peptide are in the 60S P site. The anticodon of the deacylated tRNA remains in the 40S P site, and its acceptor end is located in the 60S exit or E site.

The mRNA and the dipeptidyl-tRNA at the 40S A site must now be repositioned to permit another elongation cycle to begin. This is done by elongation factor 2 (EF-2), also called translocase. EF-2 moves the messenger and dipeptidyl-tRNA, in codon-anticodon register, from the 40S A site to the P site. In the process, GTP is hydrolyzed to GDP plus Pi, providing energy for the movement, and the A site is fully vacated. As the dipeptidyl-tRNA is moved to the P site, the deacylated donor (methionine) tRNA is also moved to the E site, which only exists on the 60S subunit. The ribosome can now enter a new cycle. The next aminoacyl-tRNA specified by the mRNA is delivered by EF-1α to the A site and the deacylated tRNA in the E site is probably released. Peptide transfer again occurs. Successive cycles of binding of aminoacyl-tRNA, peptide bond formation, and translocation result in the stepwise elongation of the polypeptide toward its eventual carboxyl terminus. Note that whatever the length of the growing chain, peptide bond formation always occurs through attack of the a-amino group of the incoming aminoacyl-tRNA on the peptide carboxyl-tRNA linkage; hence the geometric arrangement of the reacting molecules at the peptidyltransferase site remains constant. Peptide bond formation does not require any additional energy source such as ATP or GTP.

The energy of the methionyl (or peptidyl) ester linkage to tRNA drives the reaction toward peptide bond formation; recall that ATP is used to form each aminoacyl-tRNA and that these reactions are reversible. Isolated 60S subunits can catalyze peptidyltransferase activity, and nonribosomal factors are not involved in the reaction. Yet peptidyltransferase has never been dissociated from the large subunit or identified as a specific ribosomal protein. Reconstitution of E. coli peptidyltransferase activity requires only five to six different large subunit proteins and the rRNA. Omission or significant modification of the rRNA or any of these proteins causes the loss of peptidyltransferase activity, while other proteins can be deleted with little or no effect.

The discovery of catalytic RNA molecules led to speculation that the primordial ribosome was an RNA particle in which peptide bond formation was catalyzed by the RNA. Experiments with very conformationally "stable" large subunit RNA from a thermophilic bacterium suggest that the rRNA may be the catalytic component of peptidyltransferase, while the proteins serve to stabilize RNA folding; however, this hypothesis remains controversial and not fully proved. As determined with their prokaryotic equivalents, the role of GTP in the action of EF-1a and EF-2 probably relates to conformational changes in these proteins. Crystallographic studies have shown that a large rearrangement of domains with movements of several angstroms occurs upon GTP hydrolysis in EF-Tu, the prokaryotic equivalent of EF-1α. Both EF-1a and EF-2 bind ribosomes tightly as GTP complexes, while GDP complexes dissociate from the ribosome more easily. Viewed another way, GTP stabilizes a protein conformation that confers upon EF-1α high affinity toward aminoacyl-tRNA and the ribosome, while GDP stabilizes a conformation with lower affinity for aminoacyl-tRNA and ribosome, thus allowing tRNA delivery and factor dissociation. Restoration of the higher affinity GTP-associated conformation of EF-1α requires participation EF-1 βγ. This protein displaces GDP from EF-1α, forming an EF-1α · EF-1βγ complex. GTP then displaces EF-1βγ, forming an EF-1α · GTP complex that can successively bind an aminoacyl-tRNA and then a ribosome.

Termination of Polypeptide Synthesis

Termination of Polypeptide Synthesis Requires a Stop Codon A chain-terminating UAG, UAA, or UGA codon in the A site does not promote binding of any tRNA species. Instead, another complex nonribosomal protein, release factor (eRF), binds the ribosome as an eRF · GTP complex. The peptide-tRNA ester linkage is cleaved through the action of peptidyl transferase, acting here as a hydrolase, and the completed polypeptide is released from its carrier tRNA and the ribosome.

Dissociation of eRF from the ribosome requires hydrolysis of the GTP and frees the ribosome to dissociate into subunits and then reenter the protein synthesis cycle at the initiation stage. In prokaryotes three release factors, RF-1, RF-2, and RF-3, carry out the termination function. The factor RF-1 acts in response to UAG or UAA codons, RF-2 acts in response to UGA or UAA codons, and RF-3 is a GTPase that activates RF-1 and RF-2.

Translation has Significant Energy Cost

There is a considerable use of energy in synthesis of a polypeptide. Amino acid activation converts an ATP to AMP and pyrophosphate, which is normally hydrolyzed to P_i; the net cost is two high-energy phosphates. Two more highenergy bonds are hydrolyzed in the actions of EF-1α and EF-2, for a total of four per peptide bond formed. Posttranslational modifications may add to the energy cost, and of course energy is needed for biosynthesis of the multi-use mRNA, tRNAs, ribosomes, and protein factors, but these costs are distributed among the proteins formed during their lifetime.

Protein Synthesis in Mitochondria

Many characteristics of mitochondria suggest that they are descendants of aerobic prokaryotes that invaded and set up a symbiotic relationship within a eukaryotic cell. Some of their independence and prokaryotic character are retained. Human mitochondria have a circular DNA genome of 16,569 base pairs that encodes 13 proteins, 22 tRNA species, and two mitochondrion-specific rRNA species. Their independent apparatus for protein synthesis includes RNA polymerase, aminoacyl-tRNA synthetases, tRNAs, and ribosomes. Although the course of protein biosynthesis in mitochondria is like that in the cytosol, some details are different.

The synthetic components, tRNAs, aminoacyl-tRNA synthetases, and ribosomes, are unique to the mitochondrion. The number of tRNA species is small and the genetic code is slightly different. Mitochondrial ribosomes are smaller and the rRNAs are shorter than those of either the eukaryotic cytosol or of prokaryotes. An initiator Met-tRNA$^{Met}_{i}$ is modified by a transformylase that uses N^{10}-formyl H_4-folate to produce fMet-tRNA$^{Met}_{i}$. Most mitochondrial proteins are encoded in nuclear DNA and synthesized in the cytosol, but mitochondrial protein synthesis is clearly important. Cells must also coordinate protein synthesis within mitochondria with the cytosolic synthesis of proteins destined for import into mitochondria.

Toxins Inhibit Protein Biosynthesis

Protein biosynthesis is central to the continuing life and reproduction of cells. An organism can gain a biological advantage by interfering in the ability of its competitors to synthesize proteins, and many antibiotics and toxins function in this way. Some are selective for prokaryotic rather than eukaryotic protein synthesis and so are extremely useful in clinical practice. Examples of antibiotic action are listed in Table 5.

Table 5

Some Inhibitors of Protein Biosynthesis

Inhibitor	*Processes Affected*	*Site of Action*
Streptomycin	Initiation, elongation	Prokaryotes: 30S subunit
Neomycins	Translation	Prokaryotes: multiple sites
Tetracyclines	Aminoacyl-tRNA binding	30S or 40S subunits
Puromycin	Peptide transfer	70S or 80S ribosomes
Erythromycin	Translocation	Prokaryotes: 50S subunit
Fusidic acid	Translocation	Prokaryotes: EF-G
Cycloheximide	Elongation	Eukaryotes: 80S ribosomes
Ricin	Multiple	Eukaryotes: 60S subunit

Several mechanisms of interfering in ribosome subunit-tRNA interactions are utilized by different antibiotics. Streptomycin binds the small subunit of prokaryotic ribosomes, interferes with the initiation of protein synthesis, and causes misreading of mRNA. Although streptomycin does not directly bind ribosomal protein S12 of the small subunit, mutations in this protein or in the small subunit rRNA can confer resistance to or even dependence on streptomycin.

Protein S12 is involved in tRNA binding, and streptomycin alters the interactions of tRNA with the ribosomal subunit and mRNA, probably by affecting subunit conformation. Other aminoglycoside antibiotics, such as the neomycins or gentamicins, also cause mistranslation; they interact with the small ribosomal subunit, but at sites that differ from that for streptomycin. The aminoglycoside kasugamycin binds small subunits and inhibits the initiation of translation. Kasugamycin sensitivity depends on base methylation that normally occurs on two adjacent adenine moieties of small subunit rRNA. Tetracyclines bind directly to ribosomes and interfere in aminoacyl-tRNA binding. Other antibiotics interfere with elongation. Puromycin resembles an aminoacyl-tRNA; it binds at the ribosomal A site and acts as an acceptor in the peptidyltransferase reaction. However, since it does not interact with mRNA it cannot be translocated, and since its aminoacyl derivative is not in an ester linkage to the nucleoside it cannot serve as a peptide donor. Thus puromycin prematurely terminates translation, leading to release of peptidylpuromycin.

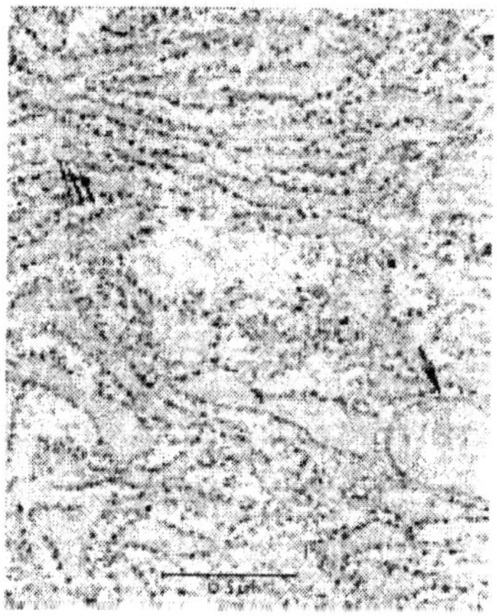

Figure.5. Rough endoplasmic reticulum of a plasma cell.

Chloramphenicol directly inhibits peptidyltransferase by binding the transferase center; no transfer occurs, and peptidyl-tRNA remains associated subunit with the ribosome. The translocation step is also a potential target. Erythromycin, a macrolide antibiotic, interferes with translocation on prokaryotic ribosomes. Eukaryotic translocation is inhibited by diphtheria toxin, a protein toxin produced by Corynebacterium diphtheriae; the toxin binds at the cell membrane and a subunit enters the cytoplasm and catalyzes the ADP-ribosylation and inactivation of EF-2, as represented in the reaction:

$$\underset{\text{(active)}}{\text{EF-2}} + \text{NAD}^+ \rightleftarrows \text{ADP-ribosyl } \underset{\text{(inactive)}}{\text{EF-2}} + \text{nicotinamide} + \text{H}^+$$

ADP-ribose is attached to EF-2 at a posttranslationally modified histidine residue known as diphthamide. Posttranslational events are discussed in the next section. A third group of toxins attack the rRNA. Ricin (from castor beans) and related toxins are N-glycosidases that cleave a single adenine from the large subunit rRNA backbone. The ribosome is inactivated by this apparently minor damage. A fungal toxin, α-sarcin, cleaves large subunit rRNA at a single site and similarly inactivates the ribosome. Some *E. coli* strains make extracellular toxins that affect other bacteria. One of these, colicin E3, is a ribonuclease that cleaves 16S RNA near the mRNA-binding sequence and decoding region; it thus inactivates the small subunit and halts protein synthesis in competitors of the colicin-producing cell.

Protein Maturation

Some proteins emerge from the ribosome ready to function, while others undergo a variety of posttranslational modifications. These alterations may result in conversion to a functional form, direction to a specific subcellular compartment, secretion from the cell, or an alteration in activity or stability. Information that determines the posttranslational fate of a protein resides in its structure: that is, the amino acid sequence and conformation of the polypeptide determine whether a protein will be a substrate for a modifying enzyme and/ or identify it for direction to a subcellular or extracellular location.

Proteins for Export Follow the Secretory Pathway

Proteins destined for export are synthesized on membrane-bound ribosomes of the rough endoplasmic reticulum (ER)(inactive). A ribosome has no means of classifying the polypeptide it is about to synthesize, so initiation and elongation begin on free cytosolic ribosomes. Proteins of the secretory pathway have a hydrophobic signal peptide, usually at or near their amino terminus. There is no unique signal peptide sequence, but its characteristics include a positively charged N terminus, a core of 8-12 hydrophobic amino acids, and a more polar C-terminal segment that eventually serves as a cleavage site for excision of the signal peptide.

The signal peptide of 15-30 amino acids emerges from the ribosome early during polypeptide synthesis. As it appears it is bound by a cytosolic signal recognition particle (SRP). The SRP is an elongated particle made up of six different proteins plus a small (7S) RNA molecule that serves as a backbone. Binding to SRP halts protein synthesis and the ribosome moves to the ER. SRP recognizes and binds to an SRP receptor or "docking protein," localized at the cytosolic surface of the ER membrane, in a reaction that requires GTP hydrolysis and presumably involves conformational changes in the SRP and/or the receptor.

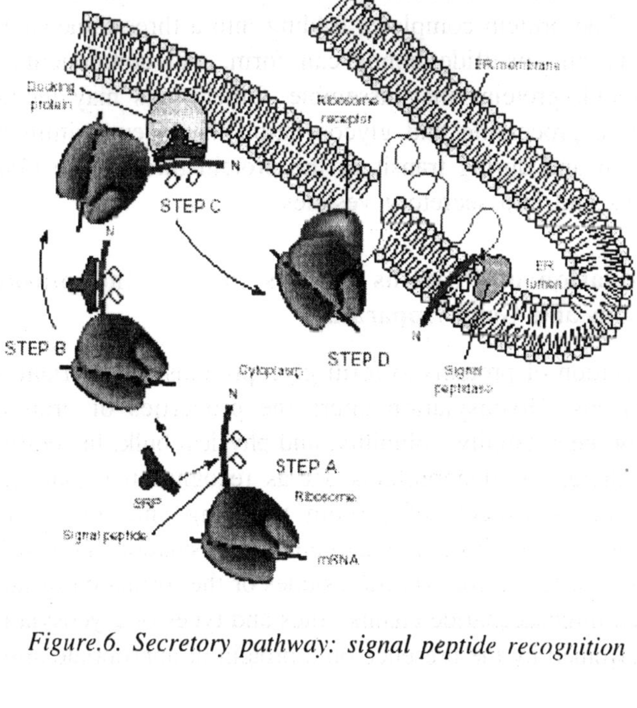

Figure.6. Secretory pathway: signal peptide recognition

The ribosome is transferred to a "translocon," a ribosome receptor on the membrane that serves as a passageway through the membrane. Both SRP and docking protein are freed to direct other ribosomes to the ER, and the translational block caused by SRP binding is relieved. The hydrophobic signal sequence, probably complexed by a receptor protein, is inserted into the membrane, further anchoring the ribosome to the ER. Translation and extrusion into or through the membrane are now coupled. Translocon proteins form a pore or channel through which the growing polypeptide passes; even very hydrophilic or ionic segments are directed through the hydrophobic membrane into the ER lumen and folding into secondary and tertiary structures begins.

The completed export-destined protein within the ER lumen will probably be anchored to the membrane by the signal peptide. A cleavage site on the protein is hydrolyzed by signal peptidase, an integral membrane protein located at the luminal surface of the ER. The protein completes folding into a three-dimensional conformation, disulfide bonds can form, and components of multisubunit proteins may assemble. Other steps may include proteolytic processing and glycosylation that occur within the ER lumen and during transit of the protein through the Golgi apparatus and into secretory vesicles.

Glycosylation of Proteins Occurs in the Endoplasmic Reticulum and Golgi Apparatus

Glycosylation of proteins to form glycoproteins is important for two reasons. Glycosylation alters the properties of proteins, changing their stability, solubility, and physical bulk. In addition, carbohydrates of glycoproteins act as recognition signals that are central to aspects of protein targeting and for cellular recognition of proteins and other cells. Glycosylation can involve addition of a few carbohydrate residues or the formation of large branched oligosaccharide chains. Sites and types of glycosylation are determined by the presence on a protein of appropriate amino acids and sequences, and by availability of enzymes and substrates to carry out the glycosylation reactions.

Glycosylation involves many glycosyltransferases, classes of which are summarized in Table 6. Up to 100 different enzymes each carry out a similar basic reaction in which a sugar is transferred from an activated donor substrate to an acceptor, usually another sugar residue that is part of an oligosaccharide under construction. The enzymes show three kinds of specificity: for the monosaccharide that is transferred, for structure and sequence of the acceptor molecule, and for the site and configuration of the anomeric linkage formed. One class of glycoproteins has sugars linked through the amide nitrogen of asparagine residues in the process of N-linked glycosylation.

The antibiotic tunicamycin, which prevents N-glycosylation, has been valuable in elucidating the biosynthetic pathway. Formation of N-linked oligosaccharides begins in the ER lumen and continues after transport of the protein to the Golgi apparatus. A specific sequence, Asn-X-Thr (or Ser) in which X may be any amino acid except proline or aspartic acid, is required for N-glycosylation. Not all Asn-XThr/ Ser sequences are glycosylated because some may be unavailable due to protein conformation.

Biosynthesis of N-linked oligosaccharides begins with the synthesis of a lipid-linked intermediate. Dolichol phosphate at the cytoplasmic surface of the ER membrane serves as glycosyl acceptor of N-acetylglucosamine. The GlcNAc-pyrophosphoryldolichol is an acceptor for stepwise glycosylation and formation of a branched $(Man)_5(GlcNAc)_2$-pyrophosphoryldolichol on the cytosolic side of the membrane. This intermediate is then reoriented to the luminal surface of the ER membrane, and four additional mannose and then three glucose residues are sequentially added to complete the structure. The complete oligosaccharide is then transferred from its dolichol carrier to an asparagine residue of the polypeptide as it emerges into the ER lumen. Thus N-glycosylation is cotranslational, that is, occurs as the protein is being synthesized, hence it can affect

Table 6. Glycosyltransferases in Eukaryotic Cells

Sugar transferred	*Abbreviation*	*Donors*	*Glycosyltransferase*
Mannose	Man	GDP-Man	Mannosyltransferase
		Dolichol-Man	
Galactose	Gal	UDP-Gal	Galactosyltransferase
Glucose	Glc	UDP-Glc	Glucosyltransferase
		Dolichol-Glc	
Fucose	Fuc	GDP-Fuc	Fucosyltransferase
N-Acetylgalactosamine	GalNAc	UDP-GalNac	N-acetylgalactosaminyltransferase
N-Acetylglucosamine	GlcNAc	UDP-GlcNAc	N-acetylglucosaminyltransferase
N-Acetylneuraminic acid	NANA or NeuNAc	CMP-NANA	N-Acetylneuraminyltransferase
(or sialic acid)	SA	CMP-SA	(sialyltransferase)

protein folding. Processing or modification of the oligosaccharide by glycosidases involves removal of some sugar residues from the newly transferred structure. The glucose residues, which were required for transfer of the oligosaccharide from the dolichol carrier, are sequentially removed, as is one mannose. These alterations mark the glycoprotein for transport to the Golgi apparatus where further trimming by glycosidases may occur. Additional sugars may also be added by a variety of glycosyltransferases. The resulting N-linked oligosaccharides are diverse, but two classes are distinguishable. Each has a common core region ($GlcNAc_2Man_3$) linked to asparagine and originating from the dolichollinked intermediate. The high-mannose type includes mannose residues in a variety of linkages and shows less processing from the dolichol-linked intermediate. The complex type is more highly processed and diverse, with a larger variety of sugars and linkages. Examples of mature oligosaccharides are shown in Figure 7.

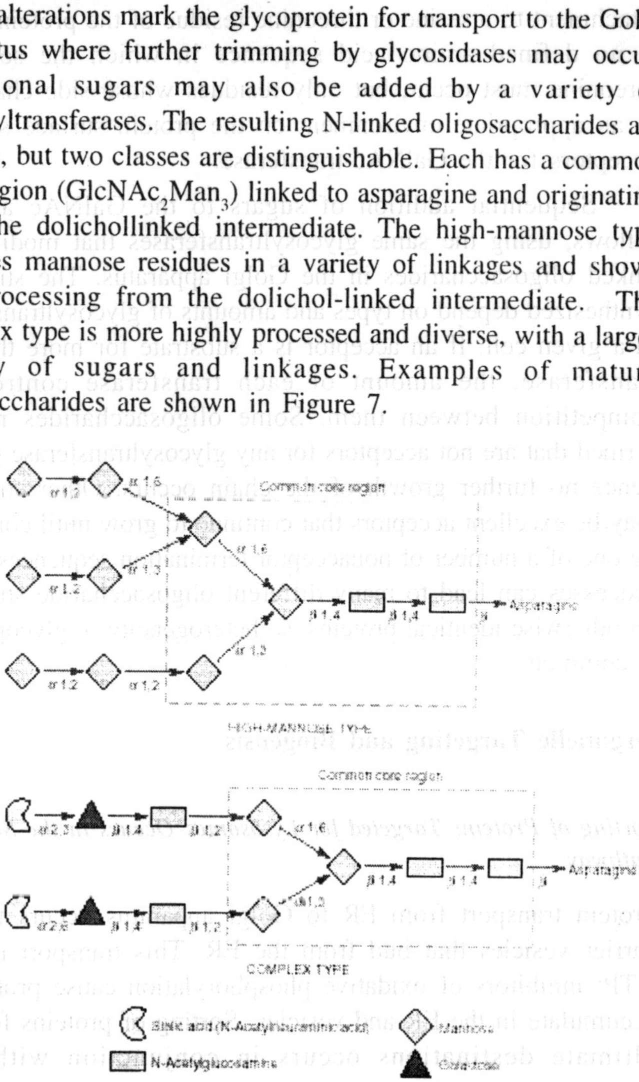

Figure.7. Structure of N-linked oligosaccharides.

The second major class of glycoproteins have sugars that are bound through either serine or threonine hydroxyl groups. Such O-linked glycosylation occurs only after the protein has reached the Golgi apparatus, hence O-glycosylation is posttranslational and occurs only on fully folded proteins. O-linked carbohydrates always involve N-acetylgalactosamine attachment to a serine or threonine residue of the protein. There is no defined amino acid sequence in which the serine or threonine must occur, but only residues whose side chains are in an appropriate environment on the protein surface serve as acceptors for the GalNAc-transferase.

Sequential addition of sugars to the GalNAc acceptor follows, using the same glycosyltransferases that modified N-linked oligosaccharides in the Golgi apparatus. The structures synthesized depend on types and amounts of glycosyltransferases in a given cell. If an acceptor is a substrate for more than one transferase, the amount of each transferase controls the competition between them. Some oligosaccharides may be formed that are not acceptors for any glycosyltransferase present, hence no further growth of the chain occurs. Other structures may be excellent acceptors that continue to grow until completed by one of a number of nonacceptor termination sequences. These processes can lead to many different oligosaccharide structures on otherwise identical proteins, so heterogeneity in glycoproteins is common.

Organelle Targeting and Biogensis

Sorting of Proteins Targeted for Lysosomes Occurs in the Secretory Pathway

Protein transport from ER to Golgi apparatus occurs through carrier vesicles that bud from the ER. This transport requires GTP; inhibitors of oxidative phosphorylation cause proteins to accumulate in the ER and vesicles. Sorting of proteins for their ultimate destinations occurs in conjunction with their glycosylation and proteolytic trimming as they pass through the cis, medial, and trans elements of the Golgi apparatus.

The best understood sorting process is targeting of specific glycoproteins to lysosomes. In the cis Golgi some aspect of tertiary structure allows lysosomal proteins to be recognized by a glycosyltransferase that attaches N-acetylglucosamine phosphate (GlcNAc-P) to high-mannose type oligosaccharides. A glycosidase then removes the GlcNAc, forming an oligosaccharide that contains mannose 6-phosphate that is recognized by a receptor protein responsible for compartmentation and vesicular transport of these proteins to lysosomes. Other oligosaccharide chains on the proteins may be further processed to form complex type structures, but the mannose 6-phosphate determines the lysosomal destination of these proteins. Patients with I-cell disease lack the GlcNAc-P glycosyltransferase and cannot correctly mark lysosomal enzymes for their destination. Thus the enzymes are secreted from the cell.

Other sorting signals are reasonably well understood. Proteins are retained in the ER lumen in response to a C-terminal KDEL (Lys-Asp-Glu-Leu) sequence, and a different sequence in an exposed C terminus signals retention in the ER membrane. Transmembrane domains have been identified that result in retention in the Golgi. Polypeptide-specific glycosylation and sulfation of some glycoprotein hormones in the anterior pituitary mediate their sorting into storage granules. Polysialic acid modification of a neural cell adhesion protein appears to be both specific to the protein and regulated developmentally. Many other sorting signals must still be deciphered to explain fully how the Golgi apparatus directs proteins to its own subcompartments, various storage and secretory granules, and specific elements of the plasma membrane. The secretory pathway directs proteins to lysosomes, the plasma membrane, or outside the cell. Proteins of the ER and Golgi apparatus are targeted through partial use of the pathway. For example, localization of proteins on either side of or spanning the ER membrane can utilize the signal recogni-tion particle in slightly different ways. If the signal sequence is downstream from the amino terminus of the protein, the amino end may not be inserted into the membrane and may remain on the cytoplasmic surface.

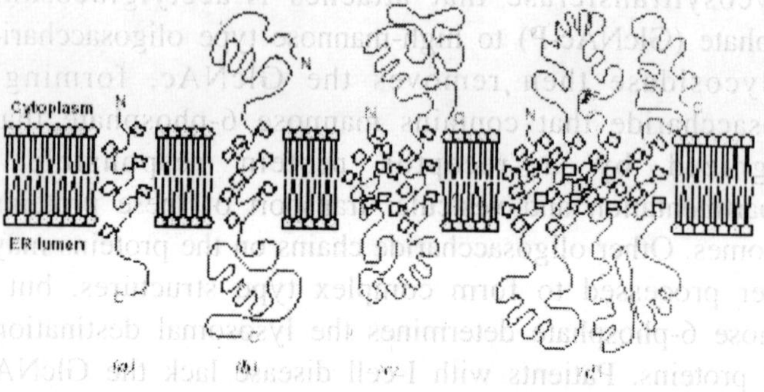

Figure.8. Topology of proteins at membranes of endoplasmic reticulum.

Internal hydrophobic anchoring sequences within a protein can allow much of the sequence either to remain on the cytoplasmic surface or to be retained, anchored on the luminal surface of the ER membrane. Multiple anchoring sequences in a single polypeptide can cause it to span the membrane several times and thus be largely buried in it. Such hydrophobic sequences are separated by polar loops whose orientation is determined by positively charged flanking residues that predominate on the cytoplasmic side of the membrane.

Import of Proteins by Mitochondria Requires Specific Signals

Mitochondria provide a particularly complex targeting problem since specific proteins are located in the mitochondrial matrix, inner or outer membrane, or intermembrane space. Most of these proteins are synthesized in the cytosol on free ribosomes and imported into the mitochondrion, and most are synthesized as larger preproteins; N-terminal presequences mark the protein not only for the mitochondrion but also for a specific subcompartment. The mitochondrial matrix targeting signal is not a specific sequence, but rather a positively charged amphiphilic a-helix. With the aid of a protein chaperone, it is recognized by a mitochondrial receptor and the protein is translocated across both membranes and into the mitochondrial matrix in an energy-dependent reaction. Passage occurs at adhesion sites where the inner and outer membranes are close together. Proteases remove the matrix targeting signal but may leave other sequences that further sort the protein within the mitochondrion.

For example, a clipped precursor of cytochromeb_2 is moved back across the inner membrane in response to a hydrophobic signal sequence. Further proteolysis frees the protein in the intermembrane space. In contrast, cytochrome-c apoprotein (without heme) binds at the outer membrane and is passed into the intermembrane space. There it acquires its heme and undergoes a conformational change that prevents return to the cytosol. Outer membrane localization can utilize the matrix targeting mechanism to translocate part of the protein, but a large apolar sequence blocks full transfer and leaves a membrane-bound protein with a C-terminal domain on the surface of the mitochondrion.

Targeting to Other Organelles Requires Specific Signals

Nuclei must import many proteins involved in their own structure and for DNA replication, transcription, and ribosome biogenesis. Nuclear pores permit the passage of small proteins, but larger proteins are targeted by nuclear localization signals that include clusters of basic amino acids. Some nuclear proteins may be retained in the nucleus by forming complexes within the organelle. Peroxisomes contain a limited array of enzymes. One targeting signal is a carboxy-terminal tripeptide, Ser-Lys-Leu (SKL). An N-terminal targeting signal also exists, and others may yet be discovered. A different targeting problem exists for proteins that reside in more than one subcellular compartment. Sometimes gene duplication and divergence have resulted in different targeting signals on closely related mature polypeptides. Alternative transcription initiation sites or pre-mRNA splicing can generate different messages from a single gene. An example of the latter is seen in a calcium-calmodulin-dependent protein kinase; alternatively spliced mRNAs differ with respect to an internal segment that encodes a nuclear localization signal. Without this segment, the protein remains in the cytosol. Alternative translation initiation sites lead to two forms of rat liver fumarase, one of which includes a mitochondrial targeting sequence while the other does not and remains in the cytosol. A suboptimal localization signal can lead to inefficient targeting and a dual location, as is seen in the partial secretion of an inhibitor of the plasminogen activator. Finally, some proteins contain more than one targeting signal, which must compete with each other.

Post-translational Protein Modifications

Several additional maturation events may modify newly synthesized polypeptides to help generate their final, functional structures. Many of these events are very common, while others are specialized to one or a few known instances.

Insulin Biosynthesis Involves Partial Proteolysis

Partial proteolysis of proteins is a common maturation step. Sequences can be removed from either end or from within the protein. Proteolysis in the ER and Golgi apparatus helps to mature the protein hormone insulin. Preproinsulin encoded by mRNA is inserted into the ER lumen. Signal peptidase cleaves the signal peptide to generate proinsulin, which folds to form the correct disulfide linkages. Proinsulin is transported to the Golgi apparatus where it is packaged into secretory granules. An internal connecting peptide (C peptide) is removed by

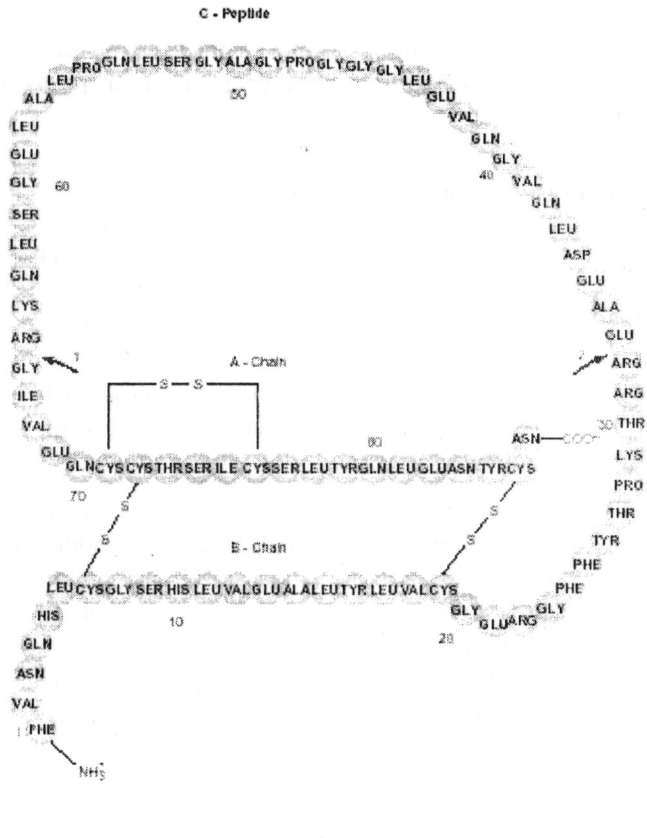

Figure.9. Maturation of human proinsulin.

proteolysis, and mature insulin is secreted. In familial hyperproinsulinemia, processing is incomplete. This pathway for insulin biosynthesis has advantages over synthesis and binding of two separate polypeptides. First, it ensures production of equal amounts of A and B chains without coordination of two translational activities. Second, proinsulin folds into a three-dimensional structure in which the cysteine residues are placed for correct disulfide bond formation. Proinsulin can be reduced and denatured but refolds correctly to form proinsulin. Renaturation of reduced and denatured insulin is less efficient, and incorrect disulfide linkages are also formed. Correct formation of insulin from separately synthesized chains might have required evolution of a helper protein or molecular chaperone.

Proteolysis Leads to Zymogen Activation

Precursor protein cleavage is a common means of enzyme activation. Digestive proteases are classic examples of this phenomenon. Inactive zymogen precursors are packaged in storage granules and activated by proteol-ysis upon secretion. Thus trypsinogen is cleaved to give an amino-terminal peptide plus trypsin, and chymotrypsinogen is cleaved to form chymotrypsin and two peptides.

Only 20 amino acids are encoded genetically and incorporated during translation. Posttranslational modification of proteins, however, leads to formation of 100 or more different amino acid derivatives in proteins. Modification may be permanent or highly reversible. The amounts of modified amino acids may be small, but they often play a major functional role in proteins. Examples are listed in Table 7.

Table 7. Modified Amino Acids in Proteins

Amino Acid	*Modifications Found*
Amino terminus	Formylation, acetylation, aminoacylation, myristoylation, glycosylation
Carboxyl terminus	Methylation, glycosyl-phosphatidylinositol anchor formation, ADP-ribosylation
Arginine	N-Methylation, ADP-ribosylation
Asparagine	N-Glycosylation, N-methylation, deamidation
Aspartic acid	Methylation, phosphorylation, hydroxylation
Cysteine	Cystine formation, selenocysteine formation, palmitoylation,
linkage to heme,	S-glycosylation, prenylation
Glutamic acid	Methylation, c-carboxylation, ADP-ribosylation
Glutamine	Deamidation, cross-linking, pyroglutamate formation

Histidine	Methylation, phosphorylation, diphthamide formation, ADP-ribosylation
Lysine	N-acetylation, N-methylation, oxidation, hydroxylation, crosslinking, ubiquitination, allysine formation
Methionine	Sulfoxide formation
Phenylalanine	β-Hydroxylation and glycosylation
Proline	Hydroxylation, glycosylation
Serine	Phosphorylation, glycosylation, acetylation
Threonine	Phosphorylation, glycosylation, methylation
Tryptophan	β-Hydroxylation, dione formation
Tyrosine	Phosphorylation, iodination, adenylation, sulfonylation, hydroxylation

Protein amino termini are frequently modified. Protein synthesis is initiated using methionine, but in the majority of proteins the amino-terminal residue is not methionine; proteolysis has occurred. The amino terminus is then sometimes modified by, for example, acetylation or myristoylation. Amino-terminal glutamine residues spontaneously cyclize; one possible result is the stabilization of the protein. Amino terminal sequences are occasionally lengthened by the addition of an amino acid. Posttranslational disulfide bond formation is catalyzed by a disulfide isomerase.

The cystine-containing protein is conformationally stabilized. Disulfide formation can prevent unfolding of proteins and their passage across membranes, so it also becomes a means of localization. As seen in the case of insulin, disulfide bonds can covalently link separate polypeptides and be necessary for biological function. Cysteine modification also occurs; multiple sulfatase deficiency arises from reduced ability to carry out a posttranslational modification. Methylation of lysine amino groups occurs in histone proteins and may modulate their interactions with DNA. A fraction of the H_2A histone is also modified through isopeptide linkage of a small protein, ubiquitin, from its Cterminal glycine to a lysine α-amino group on the histone. A role in DNA interactions is postulated. Biotin is also linked to proteins through amide linkages to lysine. Serine and threonine hydroxyl groups are major sites of glycosylation and of reversible phosphorylation by protein kinases and protein phosphatases. A classic example of phosphorylation of a serine residue is glycogen phosphorylase, which is modified by phosphorylase kinase. Tyrosine kinase activity is a property of many growth factor receptors; growth factor binding stimulates cell division and the proliferation of specific cell types. Oncogenes, responsible in part for the proliferation of tumor cells, often have tyrosine kinase activity and show strong homology with normal growth factor receptors. Dozens of other examples exist; together the protein kinases and protein phosphatases control the activity of many proteins that are central to normal and abnormal cellular development.

ADP-ribosylation of EF-2 at a modified histidine residue represents a doubling of posttranslational modifications. First, a specific EF-2 histidine residue is modified to generate the diphthamide derivative of the functional protein. This modification is probably not absolutely required since yeast mutants that cannot make diphthamide survive. ADP-ribosylation of the diphthamide by diphtheria toxin then inhibits EF-2 activity.

Other instances of physiological ADP-ribosylation not mediated by bacterial toxins are reversible. Formation of c-carboxyglutamate from glutamic acid residues occurs in several blood-clotting proteins including prothrombin and factors VII, IX, and X. The c-carboxyglutamate residues chelate calcium ion, which is required for normal blood clotting. In each case the modification requires vitamin K and can be blocked by coumarin derivatives, which antagonize vitamin K. As a result, the rate of coagulation is greatly decreased.

Collagen Biosynthesis Requires Many Posttranslational Modifications

Collagen, the most abundant protein (or family of related proteins) in the human body, is a fibrous protein that provides the structural framework for tissues and organs. It undergoes a wide variety of posttranslational modifications that directly affect its structure and function, and defects in its modification result in serious diseases. Collagen is an excellent example of the importance of posttranslational modification.

Different species of collagen, designated types I, II, III, IV, and so on are encoded on several chromosomes and expressed in different tissues. Their amino acid sequences differ, but their overall structural similarity suggests a common evolutionary origin. Each collagen polypeptide, designated an a chain, has a repeating sequence Gly-X-Y that is about 1000 residues long. Every third residue is glycine, about one-third of the X positions are occupied by proline and a similar number of Y positions are 4-hydroxyproline, a posttranslationally modified form of proline. Proline and hydroxyproline residues impart considerable rigidity to the structure, which exists as a polyproline type II helix. A collagen molecule includes three a chains intertwined in a collagen triple helix in which the glycine residues occupy the center of the structure.

Procollagen Formation in the Endoplasmic Reticulum and Golgi Apparatus

Collagen a chain synthesis starts in the cytosol, where the amino-terminal signal sequences bind signal recognition particles Precursor forms, designated, for example, prepro a1(I), are extruded into the ER lumen and the signal peptides are cleaved. Hydroxylation of proline and lysine residues occurs cotranslationally, before assembly of a triple helix. Prolyl 4-hydroxylase requires an -X-Pro-Gly- sequence (hence 4-hydroxyproline is found only at Y positions in the - Gly-X-Y- sequence). Also present in the ER is a prolyl 3-hydroxylase, which modifies a smaller number of proline residues, and a lysyl hydroxylase, which modifies some of the Y-position lysine residues. These hydroxylases require Fe^{2+} and ascorbic acid, the extent of modification depending on the specific α-chain type. Proline hydroxylation stabilizes collagen and lysine hydroxylation provides sites for interchain cross-linking and for glycosylation by specific glycosyl transferases of the ER. Asparagine residues are also glycosylated at this point, eventually leading to high mannose-type oligosaccharides.

Triple helix assembly occurs after the polypeptide chains have been completed. Carboxy-terminal globular proprotein domains fold and disulfide bonds are formed. Interaction of these domains initiates winding of the triple helix from the carboxyl end toward the amino terminus. The completed triple helix, with globular proprotein domains at each end, moves to the Golgi apparatus where oligosaccharides are processed and matured. Sometimes tyrosine residues are modified by sulfation and some serines are phosphorylated. The completed procollagen is then released from the cell via secretory vesicles.

Collagen Maturation

Conversion of procollagen to collagen occurs extracellularly. The amino-terminal and carboxyl-terminal propeptides are cleaved by separate proteases that may also be type specific. Concurrently, the triple helices assemble into fibrils and the collagen is stabilized by extensive cross-linking. Lysyl oxidase converts some lysine or hydroxylysine to the reactive aldehydes, allysine, or hydroxyallysine. These residues condense with each other or with lysine or hydroxylysine residues in adjacent chains to form Schiff's base and aldol cross-links. Further and less well-characterized reactions can involve other residues including histidines and can link three a chains. Defects at many of these steps are known.

Regulation of Translation

Translation requires considerable energy, and the formation of functioning proteins has significant consequences for the cell. It is logical that the process is carefully controlled, both globally and for specific proteins. The most efficient and common mechanism of regulation is at the initiation stage. The best understood means of overall regulation of translation involves the reversible phosphorylation of eIF-2a. Under conditions that include nutrient starvation, heat shock, and viral infection, eIF-2a is phosphorylated by a specific kinase. Phosphorylated eIF-2a · GDP binds tightly to eIF-2b, the guanine nucleotide exchange factor, which is present in limiting amounts. Since eIF-2b is unavailable for nucleotide exchange, no eIF-2a · GTP is available for initiation.

Phosphorylation can be catalyzed by a heme-regulated inhibitor kinase, which, in the absence of heme, is activated by autophosphorylation. This kinase is present in many cells but is best studied in reticulocytes that synthesize hemoglobin. Deficiencies in energy supply or any heme precursor activate the kinase. A related double-stranded RNA-dependent kinase is autophosphorylated and activated in response to binding of ds-RNA that results from many viral infections. Production of this kinase is also induced by interferon. Initiation factor eIF-4e (a component of the cap binding protein eIF-4f) is activated by phosphorylation in response to, for example, growth factors and is inactivated by a protein phosphatase following, for example, viral infection. These effects may be greatest in the translation of mRNAs with long, highly structured leader sequences that need to be unwound to allow identification of a translational start site.

Regulation of translation of specific genes also occurs. A clear example is the regulation by iron of synthesis of the iron-binding protein, ferritin. In the absence of iron a repressor protein binds to the iron-responsive element (IRE), a stem-loop structure in the 5' leader sequence of ferritin mRNA. This mRNA is sequestered for future use. δ-Aminolevulinic acid synthase, an enzyme of heme biosynthesis, is also regulated by a 59-IRE in its mRNA. In contrast, more ferritin receptor mRNA is needed if iron is limited; it has IREs in its 3'-nontranslated region. Binding of the repressor protein stabilizes the mRNA and prolongs its useful lifetime. Many growth-regulated mRNAs, including those for ribosomal proteins, have a polypyrimidine tract in their leader sequence. A polypyrimidine-binding protein helps regulate their translation.

Protein Degradation and Turnover

Proteins have finite lifetimes. They are subject to environmental damage such as oxidation, proteolysis, conformational denaturation, or other irreversible modifications. Equally important, cells need to change their protein complements in order to respond to different needs and situations. Specific proteins have very different lifetimes. Cells of the eye lens are not replaced and their proteins are not recycled. Hemoglobin in red blood cells lasts the life of these cells, about 120 days.

Other proteins have lifetimes measured in days, hours, or even minutes. Some blood-clotting proteins survive for only a few days, so hemophiliacs are only protected for a short period by transfusions or injections of required factors. Diabetics require insulin injections regularly since the hormone is metabolized. Metabolic enzymes vary quantitatively depending on need; for example, urea cycle enzyme levels change in response to diet. Most amino acids produced by protein degradation are recycled to synthesize new proteins but some degradation products will be excreted. In either case, proteolysis first reduces the proteins in question to peptides and eventually amino acids. Several proteolytic systems accomplish this end.

Intracellular Digestion of Some Proteins Occurs in Lysosomes

Digestive proteases such as pepsin, trypsin, chymotrypsin, and elastase hydrolyze dietary protein and have no part in intracellular protein turnover within an organism. Intracellular digestion of proteins from the extracellular environment occurs within lysosomes. Material that is impermeable to the plasma membrane is imported by endocytosis. In pinocytosis large particles, molecular aggregates, or other molecules present in the extracellular fluid are ingested by engulfment. Macrophages ingest bacteria and dead cells by this mechanism. Receptor-mediated endocytosis uses cell surface receptors to bind specific molecules. Endocytosis occurs at pits in the cell surface that are coated internally with the multisubunit protein clathrin. Uptake is by invagination of the plasma membrane and the receptors to form intracellular coated vesicles. One fate of such vesicles is fusion with a lysosome and degradation of the contents.

Some intracellular protein turnover may also occur within lysosomes, and under some conditions significant amounts of cellular material can be mobilized via lysosomes. For example, serum starvation of fibroblasts in culture or starvation of rats leads to the lysosomal degradation of a subpopulation of cellular proteins. Recognition of a specific peptide sequence is involved, indicating that the lifetime of a protein is ultimately encoded in its sequence. This concept will be more apparent in the next section on ubiquitin-dependent proteolysis. Although lysosomal degradation of cellular proteins occurs, it is not the main route of protein turnover.

Calcium-dependent proteases, also called calpains, are present in most cells. Activators and inhibitors of these enzymes are also present, and calpains are logical candidates for enzymes involved in protein turnover. However, their role in these processes is not quantitatively established. Golgi and ER proteases degrade peptide fragments that arise during maturation of proteins in the secretory pathway. They could also be involved in turnover of ER proteins. Apoptosis, programmed cell death, requires several proteases. It is likely that other uncharacterized mechanisms exist in both the cytosol and in the mitochondrion.

Ubiquitin Is a Marker in ATP-Dependent Proteolysis

One well-described proteolytic pathway requires ATP hydrolysis and the participation of ubiquitin, a highly conserved protein containing 76 amino acids. One function of ubiquitin is to mark proteins for degradation. Ubiquitin has other roles; as an example, linkage of ubiquitin to histones H2A and H2B is unrelated to turnover since the proteins are stable, but modification may affect chromatin structure or transcription. The ubiquitin-dependent proteolytic cycle is shown in Figure 10.

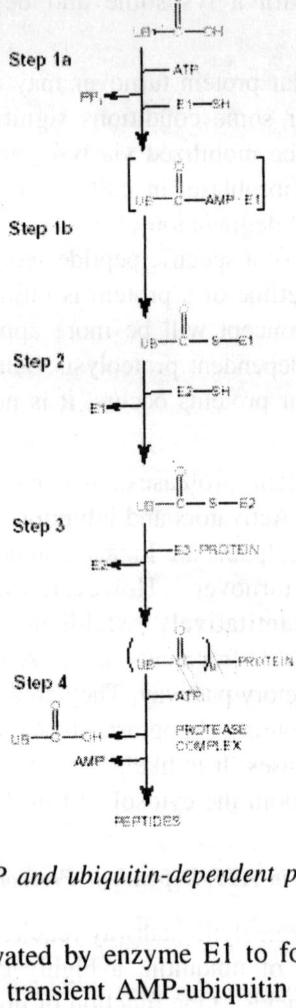

Figure.10. ATP and ubiquitin-dependent protein degradation

Ubiquitin is activated by enzyme E1 to form a thioester; ATP is required and a transient AMP-ubiquitin complex is involved. The ubiquitin is then passed to enzyme E2, and finally via one of a group of E3 enzymes it is coupled to a targeted protein.

Linkage of ubiquitin is through isopeptide bonds between amino groups of lysine residues of the protein and the carboxyl-terminal glycine residues of ubiquitin. Several ubiquitin molecules may be attached to the protein and to each other. ATP-dependent proteases then degrade the tagged protein and free the ubiquitin for further degradation cycles.

Ubiquitin-dependent proteolysis plays a major role in the regulation of cellular events. Cyclins are involved in control of progress through the cell cycle. The ubiquitin-dependent destruction of a cyclin allows cells to pass from the M phase into G1. Other proteins known to be degraded by ubiquitindependent proteolysis include transcription factors, the p53 tumor suppressor and other oncoproteins, a protein kinase, and immune system and other cell surface receptors.

Figure.11. Model of proteusome

Damaged or mutant proteins are rapidly degraded via the ubiquitin pathway. In cystic fibrosis a mutation that results in deletion of one amino acid greatly alters the stability of a protein, but it is not always clear how native proteins are identified for degradation. Selectivity occurs at the level of the E3 enzyme, but most specific recognition signals are obscure.

One determinant is simply the identity of the amino-terminal amino acid. Otherwise identical b-galactosidase proteins with different amino-terminal residues are degraded at widely differing rates. Amino termini may be modified to alter the lifetime of the protein, and some residues serve as aminoacyl acceptors for a destabilizing residue from an aminoacyl-tRNA. Internal sequences and conformation are also likely to be important; destabilizing PEST sequences (rich in Pro, Glu, Ser, and Thr) have been identified in several short-lived proteins. The ATP-dependent degradation of ubiquitin-marked proteins occurs in a 26S organelle called the proteasome. Proteasomes are dumbbell-shaped complexes of about 25 polypeptides; a proteolytically active 20S cylindrical core is capped at each end by V-shaped complexes that bestow ATP dependence. It is speculated that the cap structure is involved in recognizing and unfolding polypeptides and transporting them to the proteolytic core. The complex E. coli proteases Lon and Clp and similar enzymes in other microorganisms (and in mitochondria) also require ATP hydrolysis for their action, but ubiquitin is absent in prokaryotes and the means of identification of proteins for degradation is still obscure. It is likely that protein degradation will turn out to be as complex and important a problem as protein biosynthesis.